E. Lien

W9-CFD-517

NOV 2 7 1991

New Protein Foods in Human Health:

NUTRITION, PREVENTION, AND THERAPY

Fred H. Steinke
Doyle H. Waggle
Michel N. Volgarev

CRC Press
Boca Raton   Ann Arbor   London

Library of Congress Cataloging-in-Publication Data

New protein foods in human health : nutrition, prevention, and therapy
  / editors, Fred H. Steinke, Doyle H. Waggle, Michael N. Volgarev.
     p. cm.
  Includes bibliographical references and index.
  ISBN 0-8493-6904-5
  1. Soyfoods. 2. Proteins in human nutrition. I. Steinke, Fred
H. II. Waggle, Doyle H. III. Volgarev, Michael N.
TX558.S7N48 1991
664′.805655--dc20   91-27279
                                                       CIP

Direct all inquiries to CRC Press, Inc., 2000 Corporate Blvd., N.W., Boca Raton, Florida 33431.

© 1992 by CRC Press, Inc.

International Standard Book Number 0-8493-6904-5

Library of Congress Card Number 91-27279

Printed in the United States of America   1 2 3 4 5 6 7 8 9 0

THE EDITORS

Fred H. Steinke, Ph.D., was Director (retired), Nutritional Science, Research and Development, Protein Technologies International, Ralston Purina Company.

Dr. Steinke obtained his B.S. (1957) and M.S. (1959) degrees from the University of Delaware, and his Ph.D. degree in Biochemistry from the University of Wisconsin in 1963.

Dr. Steinke joined the Ralston Purina Company in 1963 and has worked on the nutritional value of both animal and human food products. Responsibilities have included research and development of feed products for laying hens, turkeys, and ducks. Since 1972, Dr. Steinke has been responsible for establishing nutritional information on isolated soy protein, soy fiber, and food products using these ingredients. These responsibilities required conducting extensive animal research programs and coordinating human clinical studies through grant programs with research institutes throughout the world. His responsibilities also involve communication of nutritional information to customers, scientists, and government personnel throughout the world.

Dr. Steinke is the author or co-author of over 45 scientific papers and three patents.

Doyle H. Waggle, Ph.D., is Vice President and Director of Technology, Protein Technologies International, Inc., wholly owned subsidiary of Ralston Purina Company, St. Louis, Missouri.

Dr. Waggle received his B.S. degree from Kansas State College, Ft. Hayes, Kansas in 1961. He obtained his M.S. and Ph.D. degrees in 1963 and 1966, respectively, from Kansas State University, Manhattan, Kansas.

He joined the Ralston Purina Company in 1967. His responsibilities include research and development of protein and fiber based products for the food industry.

Dr. Waggle is a member of the Industrial Research Institute, the Institute of Food Technologists, and the American Association of Cereal Chemists. He is the author or co-author of more than 20 papers and books.

Michael N. Volgarev, M.D., Ph.D., D.Sc., is Professor and Corresponding Member of the USSR Academy of Medical Sciences, and Director of the Institute of Nutrition.

Dr. Volgarev graduated (pre-med) from the First Moscow Medical Institute in 1954; in 1958 he presented a thesis and received the degree of Candidate of Medicine; and in 1970 he received the degree of Doctor of Medicine. In 1985 the academic rank of Professor of Pathologic Anatomy was conferred on him, and in 1986 he was elected a Corresponding Member, USSR Academy of Medical Sciences.

Dr. Volgarev is a member of the USSR Physiologic Society, AMS Prize Laureate on Histology. He has published approximately 140 articles and 1 monograph.

His current areas of research include RDA (recommended dietary allowance) studies, the influence of dietary disorders and food contamination upon carcinogenesis, and human immune status.

CONTRIBUTORS

ANDREI N. BOGATYREV, C.T.S., Head, Agro-Industry, GKNT, USSR Committee for Science and Technology, Moscow, USSR

TATYANA E. BOROVICK, M. D., Junior Researcher, Laboratory for Nutrition of Sick Children, Institute of Nutrition, Academy of Medical Sciences, Moscow, USSR

SAMUEL J. FOMON, M. D., Professor, Department of Pediatrics, College of Medicine, University of Iowa, Iowa City, IA, USA

SCOTT M. GRUNDY, M.D., Ph.D., Director, Center for Human Nutrition, University of Texas, Southwestern Medical Center, Dallas, TX, USA

OLGA M. KALAMKAROVA, Ph.D., M.D., Junior Research, Laboratory for Evaluation of Food Proteins, Institute of Nutrition, Academy of Medical Sciences, Moscow, USSR

CHARLES W. KOLAR, Ph.D., Associate Director, Food Science, Research and Development, Protein Technologies International, Inc., St. Louis, MO, USA

KALERIA S. LADODO, M.D., D.Sc., Professor, Head, Department for Child Nutrition, Institute of Nutrition, Academy of Medical Sciences, Moscow, USSR

BORIS G. LYAPKOV, M.D., D.Sc., Head of the Laboratory for Instrumental Methods of Analysis, Institute of Nutrition, Academy of Medical Sciences, Moscow, USSR

VALENTINA A. MESHCHERYAKOVA, M.D., Ph.D., Head of the Department of Clinical Dietetics, Institute of Nutrition, Academy of Medical Sciences, Moscow, USSR

OXANA A. PLOTNIKOVA, M.D., Ph.D., Senior Researcher, Department of Clinical Dietetics, Institute of Nutrition, Academy of Medical Sciences, Moscow, USSR

GALINA R. POKROVSKAYA, M.D., Ph.D., Senior Researcher, Department of Cardiovascular Pathology, Institute of Nutrition, Academy of Medical Sciences, Moscow, USSR

LI KHVA REN, Ph.D., Researcher, Laboratory for Enzymology of Nutrition, Institute of Nutrition, Academy of Medical Sciences, Moscow, USSR

STANLEY H. RICHERT, Ph.D., Vice President of Meat Science, Research and Development, Protein Technologies International, Inc., St. Louis, MO., USA

ANNA M. SAFRONOVA, Ph.D., Junior Researcher, Laboratory for Evaluation of Food Proteins, Institute of Nutrition, Academy of Medical Sciences, Moscow, USSR

MICHAEL A. SAMSONOV, M.D., D.Sc., Professor, Head of the Department for Cardiovascular Pathology, Institute of Nutrition, Academy of Medical Sciences, Moscow, USSR

NATALYA P. SHIMANOVSKAYA, M.D., Ph.D., Junior Researcher, Department of Cardiovascular Pathology, Institute of Nutrition, Academy of Medical Sciences, Moscow, USSR

FRED H. STEINKE, Ph.D., Director (retired), Nutritional Science, Protein Technologies International, Inc., St. Louis, MO., USA

BENJAMIN TORUN, M.D., Ph.D., Head, Metabolism and Clinical Nutrition Program, Institute of Nutrition of Central America and Panama, Guatemala City, Guatemala

VICTOR A. TUTELYAN, Professor, M.D., D.Sc., Associate Director, Institute of Nutrition, Academy of Medical Sciences, Moscow, USSR

ANDREI V. VASILYEV, Ph.D., Head, Laboratory for Clinical Biochemistry, Institute of Nutrition, Academy of Medical Sciences, Moscow, USSR

ALEXANDER S. VITOLLO, Ph.D., Senior Researcher, Laboratory for Evaluation of Food Proteins, Institute of Nutrition, Academy of Medical Sciences, Moscow, USSR

MICHAEL N. VOLGAREV, Professor, M.D., D.S., Director, Institute of Nutrition, Academy of Medical Sciences, Moscow, USSR

VADIM G. VYSOTSKY, Professor, M.D., D.Sc., Head of the Laboratory for Evaluation of Food Proteins, Institute of Nutrition, Academy of Medical Sciences, Moscow, USSR

DOYLE H. WAGGLE, Ph.D., Vice President, Research and Development, Protein Technologies International, Inc., St. Louis, MO.,USA

WALTER J. WOLF, Ph.D., Biopolymer Research, National Center for Agricultural Utilization Research, USDA-ARS, Peoria, IL., USA

TAISYA A. YATSYSHINA, M.D., Ph.D., Senior Researcher, Laboratory for Evaluation of Food Proteins, Institute of Nutrition, Academy of Medical Sciences, Moscow, USSR

VERNON R. YOUNG, Ph.D., D.Sc., Professor, School of Science, Massachusetts Institute of Technology, Cambridge, MA., USA

ANATOLI I. ZAICHENKO, M.D., Ph.D., Deputy Chief, Sanitary Inspector, Ministry of Health, Moscow, USSR

VALENTIN M. ZHMINCHENKO, M.D., Ph.D., Senior Researcher, Laboratory for Evaluation of Food Proteins, Institute of Nutrition, Academy of Medical Sciences, Moscow, USSR

EKHARD E. ZIEGLER, M.D., Professor, Department of Pediatrics, College of Medicine, University of Iowa, Iowa City, IA., USA

IRINA S. ZILOVA, M.D., Ph.D., Senior Researcher, Laboratory for Evaluation of Food Proteins, Institute of Nutrition, Academy of Medical Sciences, Moscow, USSR

TABLE OF CONTENTS

Chapter 1

INTRODUCTION

Michael N. Volgarev and Doyle H. Waggle

One of the goals of nutrition for both developing and industrialized societies is to provide the population with adequate amounts of food proteins to meet the physiological/nutritional requirements of the population.[1] The protein and amino acids needs have been more clearly identified in recent years through human evaluations of protein and amino acid requirements under a variety of conditions. The problem of adequate protein and amino acid nutrition is primarily a problem of inadequate production and distribution of high-quality food proteins. The limitations in production and distribution are related to a number of factors, including the high cost of production of some types of protein (i.e., animal protein), limited production capability, climatic/environmental factors, and historic cultural restraints. The high cost of production of good quality, acceptable proteins plays an important role in any national economic planning. Therefore, the expansion of new food protein sources represents an important strategic direction in the national economic plan.

The current concepts for increasing food protein production are based on the scientific and technological methods available in agricultural production and the food processing industry. One solution to the food protein shortage based on the present technology is to intensify the existing agricultural and processing methods to obtain incremental improvements in protein production. The amount of additional protein production obtained by this method is limited and increasingly more expensive. A second approach for increasing the supply of food protein is to develop new food protein sources from economical production sources and biotechnology which convert secondary protein sources into good-quality protein for the human diet.[2-10] The latter approach uses the resources of the biosphere more effectively and converts existing waste products into usable food proteins.[3,11,12] The direct use of food proteins by humans will eliminate the wasteful losses incurred through conversion to animal products.

The development of new food proteins from secondary and new sources in economically developed countries is needed to meet protein requirements, and is also equally important for preventing chronic diseases associated with the consumption of large amounts of animal foods. These dietary changes can be implemented rapidly in developed countries through the use of the existing technological and processing infrastructure.

This book will summarize the scientific data on the use of new protein isolates as protein and amino acid sources to augment existing food protein sources. In addition, the review will look at the potential benefits of new protein sources as they relate to the prevention and treatment of diet-related diseases which are prevalent in developed societies. The current concepts for the possible uses for new protein sources and by-products of the food industry are outlined in Table 1. A broad range of practical applications is possible for new protein sources in the nutritional field which involve the creation of a new generation of food products. Among the primary reasons for developing these new foods sources are the existence of inadequate nutrition, the inadequate and wasteful use of existing nutritional resources, and the proven need for prevention and therapeutic nutrition to optimize health and longevity.[13]

In the area of general nutrition, the broad application of various high-protein preparations (protein isolates, and to a limited extent concentrates or flours) during the manufacture of processed products and foodstuffs will make it possible to utilize the most valuable types of raw foods (e.g., meat) more judiciously without sacrificing the biological value of its

TABLE 1
Applications of Proteins From Nontraditional and New Sources

General Diets

Substitute for high-value animal proteins
Enriching low-value vegetable proteins
Control the chemical composition of foodstuffs
Develop specialized child nutrition products (substitutes for mother's milk, weaning food, products with an elevated
  biological value)

Therapeutic Diets

Develop food products having a set chemical composition for correcting metabolic disruptions or food of intolerance
Products for complete diet
Products for weight-reduction diets

Preventative Diets

Low-calorie foods
Products with lower fat and cholesterol levels
Specialized products for individuals working under physiologically stressful conditions

proteins. It will additionally permit the more effective use of vegetable protein resources based on the mutual enrichment of proteins and to develop combined food products having a controlled chemical composition. Although this is beneficial from an economic point of view, it also has great medical significance for the public from the standpoint of safeguarding the health of the population. A second area of the application for new protein sources has to do with developing formulas and technologies for manufacturing food products which aid in the dietary treatment of food allergies and intolerances. The need for these products stems not only from economic considerations, but also because new opportunities are opened for manufacturing products having controlled chemical composition and high nutritional and biological value with antinutritional and toxic substances removed to the maximum extent possible. This is particularly important with respect to high-risk groups and above all for child nutrition.

The practical opportunity of using new protein preparations in the manufacture of products in preventive diets has great medical and social importance. The most realistic way of making products having fewer calories, a lower cholesterol level, a lower fat content, or an elevated protein content is to include protein isolates, which contain virtually no lipids. Vegetable protein isolates are the most interesting in this regard since they contain no cholesterol and some of them exert a hypocholesterolemic effect.[14-16] Additional uses of new protein preparations could include special diets for individuals who work under conditions of physiological stress or heavy physical activity by providing the body with proteins having a specific composition of amino acids.

For therapeutic diets the only practical method for ensuring that these products will contain the protein component is to use protein isolates and the biomedical basis for their application. This category of products requires the development of special technologies which make it possible to obtain protein components having various physiochemical characteristics. Native proteins may in several instances be employed, depending on the nature of the illness, while protein hydrolysates must be used in other cases. In addition, preference should in each of these cases be given to proteins, hydrolysates, or combinations of these, whose essential amino acids are as balanced as possible for human requirements. From a practical standpoint, the problem of therapeutic diets hinges on the production of various kinds of protein isolates and hydrolysates.

TABLE 2
Sources of Food Protein

Traditional Protein Sources

Products from:
 Plant growing
 Animal husbandry
 Poultry farming
 Fishing and fish breeding
 Natural flora and fauna

Nontraditional Protein Sources

Extracted isolates from seeds of the following cultures:
 Oil producing (soya, sunflower, rape, ground nuts)
 Food (tomatoes, grapes, corn kernels)
 Safflower
 Fiber (cotton, flax)
 Peas, beans
Waste from milling plants and hulling mills
Biomass from leaf mass protein, nonfat milk, and whey
Blood, several tissues, and organs of animals
Noncommercial fish and sea products not used in food

New Protein Sources

Single- and multicelled alga
Mycelium of higher and lower fungi
Yeasts
Nonpathogenic bacteria
Protein from chemical synthesis

Bypassing intermediate steps in the food chains and directly using proteins in the human diet depends not only on finding solutions to technological problems, but also on the biomedical justification for using these proteins for nutritional purposes. It is necessary in the latter case to analyze the chemical composition, ascertain the safety, determine the biological value and availability, as well as evaluate the best nutritional application of new proteins. This aspect of resolving the protein problem primarily involves examining the quality of proteins obtained from nontraditional raw material and new sources. From a biomedical standpoint, the word *quality* must here be understood to refer not only to the consumer, functional, and sensory characteristics of the protein product, but also to the safety of the latter in foods (i.e., the absence of antinutritional and toxic substances in quantities dangerous to the health of the population) and to the nutritional and biological value.

The list of new sources of protein showing promise in terms of application in the human diet should include protein-containing products, such as by-products from food and fodder plants, resources sparingly utilized for nutritional purposes, and completely new protein sources. The best end products obtained from processing these raw materials are protein isolate and hydrolysates, since these forms of protein products are not only most suited for food products mentioned above,[3] but also serve as a unique way for removing toxic and antinutritional substances which might be contained in the initial raw material.[3,17-21]

The entire gamut of food protein resources currently being employed in nutrition and which viewed as a potential reserve for increasing protein resources may be divided into three main groups (Table 2): traditional, nontraditional, and new sources of protein.[9,13,22] While the classification of protein sources in the first group does not require any special elucidation, the protein-containing raw material in the second group is best divided into the following three subgroups according to their properties (Table 2):

1. Secondary protein-containing products widely used in the food industry: nonfat milk, caseinates, whey, blood, isolated soya proteins, and to a lesser extent concentrates and by-products from milling plants.
2. Sources of protein slated for use in the near future for the industrial production of food proteins, but as of yet insufficiently studied from the standpoint of evaluating whether or not all of the technology for their preparation has been fully identified. These include several types of noncommercial fish and other sea products, biomass of green plants, bacterial cultures, extracted isolates from sunflower seeds, cottonseed, peas, beans, groundnuts, safflower, and several other oil-producing varieties.
3. Protein-containing raw materials which are of practical value, but have not yet been sufficiently examined from the standpoint of quality. These include several types of sea products not employed in foods, extracted isolates from the seeds of rape and other crucifers, and several tissues and organs from animals. The third group also includes new types of protein sources which have yet to be examined with respect to the safety of the proteins: single- and multicelled alga, the mycelium of higher and lower fungi, yeasts, and nonpathogenic bacteria, as well as proteins from chemical synthesis.

Such a classification determines not only the policy with regard to evaluating which protein resources are most expedient from an economical standpoint, but also serves as the basis for selecting the directions to be taken in the area of scientific research. This means that the gradation of new protein sources presented above may act as a reference point to be drawn upon in resolving the tactical and strategic tasks of finding additional sources of protein for the population. As already indicated, protein isolates represent the most promising forms of the final product obtained from processing the initial raw material. These isolates may be obtained from any protein-containing source.

The advantages offered by these forms are the following: (1) the product obtained may be stored for a considerably longer period of time than the initial raw material; (2) the product is freed of as many of the antinutritional and toxic ingredients as possible, or at least of the amount necessary for them to be within the range of maximum permissible concentrations; and (3) the concentration of protein achieved is nearly optimal, which is a trait of some importance when employing the protein for purposes of enrichment, or creating analogs of foods or combined foodstuffs. This latter form is the most suited for developing products for enteral diets and child nutrition. It is precisely this fact which explains why all of the efforts being made to utilize sparingly employed and nontraditional sources of protein are aimed primarily at developing technologies for producing isolated forms of the protein. Their production represents the most realistic way of complying with the primary requirement placed on products recommended for consumption: to be useful and safe for human consumption. One may conclude from the above that studying the safety and nutritional value of new and supplementary sources of protein is the most important task to be accomplished while attempting to resolve the complex problem concerning the utilization of these proteins in food.

In the final analysis, the safety of protein isolates and concentrates depends on the quality of the source and the degree to which the technology for extracting them has been perfected. Every new source of food protein must be considered as a possible carrier of toxic, allergic, antinutritional, and other undesirable substances.[17-21] The potential nutritional value of these resources is the determining factor when justifying the selection of a new source of food protein isolates and concentrates. The value of the final product will be determined mainly by the biological value of the protein. This is because virtually all remaining food components in the isolates will be removed during the process of isolation, and those minimal concentration amounts remaining will not substantially affect the nutritional value.

TABLE 3
Quality Indices for Protein Products

I.   NUTRITIONAL VALUE FACTORS
     A.   Content of protein and other food substances
     B.   Amino acid composition of proteins
     C.   Biological value
     D.   Assimilability of proteins

II.   SAFETY FACTORS
     A.   Natural components of protein sources
         1.   Uncommon from quantitative or qualitative standpoint
             a.   Oligosaccharide (stachyose, raffinose, etc.)
             b.   Fatty acids (erucic, cyclopropylic, etc.)
             c.   Amino acids (diaminopimelic, diaminobuteric, etc.)
             d.   Nucleic acids
             e.   Allergens
             f.   Goitrogens
             g.   Estrogens
         2.   Antialimentary
             a.   Enzyme inhibitors
             b.   Antivitamins
             c.   Demineralizing chelates
         3.   Toxins
             a.   Fungus toxins (muscarine, amantine, etc.)
             b.   Toxins for sea products (saxitoxins, gonyautoxins, etc.)
             c.   Alkaloids (divisine, pyzzolysidine, etc.)
             d.   Glycosides
             e.   Lectin
             f.   Gossypol
     B.   Environmental contaminants
         1.   Biological
             a.   Bacteria and bacterial toxins
             b.   Fungi and mycotoxins
         2.   Chemical
             a.   Pesticides and by-products
             b.   Toxic elements
             c.   Radionuclides
             d.   Polycyclic aromatic hydrocarbons
             e.   Polychlorinated biphenyls
             f.   Hormones and other growth stimulants
             g.   Antibiotics
             h.   Tranquilizers
     C.   Substances which are introduced or formed during technological processing
         1.   Emulsifiers, antifoaming agents, surfactants, and other technological substances
         2.   Food additives
         3.   Substances which form during exposure to alkali
         4.   Compounds which form during exposure to heat or radiation

The production process cannot guarantee that the various undesirable substances whose content in the initial raw material may be quite high will be removed. Chemical, clinical, and biological examinations must be conducted to validate their safety and nutritional value. The outline in Table 3 summarizes the parameters which are important to the nutritional value and safety of protein isolates obtained from a wide variety of sources. The nutritional value of these protein preparations is determined primarily by the digestibility and content of essential amino acids or the biological value. The amino acid composition of various proteins can differ widely.[23]

The levels of various undesirable substances that may be contained in the protein isolates are also presented in Table 3 and represent safety criteria. The list of substances presented in Table 3 are not specific for all protein preparations. The qualitative and quantitative characteristics of toxic and other undesirable substances will vary as a function of the type of initial raw material or the characteristic features of the technology employed in processing it. The safety of protein preparations will be determined solely by several general (e.g., toxic elements and bacterial contaminants) and specific (e.g., trypsin inhibitor in legumes/bean/peas, gossypol in cotton seeds) kinds of contaminants.

Based on an analysis of classifications proposed earlier,[9,19-21,24] the entire list of substances and indices presented in Table 3 may be divided into three primary groups according to the criterion of their origin: those which are naturally present in sources of protein, those which come from the environment, and those which are used or formed during the process of obtaining protein preparations from raw material.[13]

Included in the second group (II.A) are normal components in raw food, whose quantities in protein isolates are either regulated by the ingredient source or from the technological procedures. The second subgroup (II.B) of chemical and biological contaminants includes contaminants from the surrounding environment and pollutants. This requires special attention with regard to the hygienic procedures used in protein preparations. The majority of the latter either exert acute or chronic toxic effects.

The third group of substances (II.C) includes a broad range of compounds which are employed as technological substances while processing raw material into protein isolates and concentrates or used as food additives during the production of food products. While data pertaining to regulations and corresponding methods for analysis are available for substances in the first two subgroups,[25] the information about substances in the third subgroup is limited and frequently contradictory in terms of standardizing the methods. Any hygienic basis for the guaranteed safety of protein products obtained from new and nontraditional sources in the human diet must therefore be supported by chemical analysis of antinutritional, toxic substances, food additives, and also on the results of experimental animal feeding-evaluation examinations.

The nontraditional nature of the raw material coupled with the novelty of the technology for obtaining new food proteins may result in the final product being insufficiently purified of undesirable substances which are either of unknown origin or were not identified by a chemical analysis. The unfavorable effect which such protein products might have should therefore be evaluated in biomedical examinations. This approach is also necessary due to the fact that not enough is currently known about the cumulative and additive effects exerted by foreign substances whose content in the food corresponds to the maximum permissible levels.

In addition to the safety issues reviewed above, the new food proteins must also pass the critical test related to consumer acceptance. These factors include texture, taste, and maintenance of traditional food characteristics in manufacturing and consumer use. Therefore, the new food protein should be evaluated in the processing technologies and the consumer food evaluated through trained and untrained sensory evaluation panels. It is a poor use of resources to develop a hygienically safe and nutritionally acceptable new protein source which will not be consumed by the public due to poor flavor, texture, odor, or appearance.

It is evident from the above that there are important medical and social problems to be resolved in the areas of socialand food hygiene. On the one hand, the structure of nutrition for the population of the country must be improved. This may be accomplished by the scientifically based application of protein preparations obtained from nontraditional raw material and new sources for nutritional purposes. On the other hand, it must grapple with the issue of ensuring the safety of protein-containing food products while at the same time ensuring that unnecessary barriers are not erected which slow down innovation and availability of new resources for feeding the population. These problems are summarized in

TABLE 4
Improving Nutrition and Ensuring the Safety of Protein Products

Standardization

Preventing and treating diseases of protein deficiency and excess energy consumption
1. Developing biomedical criteria and quality indices for protein isolates and concentrates and food products in which they are used
2. Creating protein products for mass consumption with a set quality for the primary prevention of disease in a healthy population (made low calorie by replacing fat with protein, with an elevated protein content and biological value as well as a reduced cholesterol level)
3. Developing specialized protein products for the primary prevention of disease in risk groups (children, the elderly, and pregnant and nursing women)
4. Creating specialized protein products for the secondary prevention and treatment of disease

Regulation

Preventing intoxication and unfavorable remote effects
1. Developing maximum permissible quantities, systems, and methods for monitoring the contaminant levels in various protein products
2. Regulating the content of natural and anthropogenic toxic and other kinds of contaminants in protein preparations
3. Regulating the use of pesticides in plant-growing and fodder additives, growth stimulators, and other substances in animal husbandry and monitoring the quality of production
4. Regulating the use of technological methods for isolating and concentrating proteins and using them to obtain foodstuffs
5. Biomedically evaluating the quality of new protein preparations and developing rational ways of using them

general terms in Table 4. The need for resolving the stems from the formation of a new scientific concept for improving the nutrition of the population which is aimed at increasing the role of the nutrition factor in the primary prevention of diseases and providing a healthy lifestyle.[13] It is possible to develop this concept only with the production of the new protein foods. It would also be possible to handle all the tasks involved in developing and industrially incorporating the output of the entire gamut of specialized products for children's dietetic and therapeutic foods only with this availability of new protein ingredients and foods.

REFERENCES

1. Energy and Protein Requirements, Report of Joint FAO/WHO/UNU, Expert Consultation, WHO Tech. Rep. Ser. 724, Geneva, 1985.
2. Foodstuffs and Nutrition: The Strategy Within the Framework of National Development, 9th report by the Joint FAO/WHO Committee of Experts on Nutrition, Tech. Rep. Ser. 584, Geneva, 1977, 9.
3. **Tolstoguzov, V. B.**, *Synthetic Foods*, Nauka, Moscow, 1978, 9.
4. **Pokrovsky, A. A.**, *Single Celled Protein*, a review, Main Administration of the Microbiological Industry, SONTI, Moscow, 1971, 3.
5. **Morozova, G. R., Vysotsky, V. G., Safronova, A. M., and Mamaeva, E. M.**, *Industrial Production of Mycelia from Higher Fungi*, ONTITEI, Moscow, 1978, 3.
6. **Mamaeva, E. M., Vysotsky, V. G., and Zhminchenka, V. M.**, Hygiene aspects of the possibility of employing tobacco plants as sources of food protein, *Vopr. Pitan.*, 6, 13, 1955.
7. **Vysotsky, V. G. and Zilova, I. S.**, Biomedical evaluation of the quality and promise for application of proteins from several sea products, *Bull. Sov. Acad. Med. Sci.*, 12, 10, 1986.
8. *Sources of Food Protein*, Kolos, Moscow, 1979.
9. **Hardin, C. M.**, Impact plant proteins in worldwide food systems, in *Soy Protein and Human Nutrition*, Wilcke, H. L., Hopkins, D. T., and Waggle, D. H., Eds., Academic Press, New York, 1979, 5.
10. **Ferrando, R.**, Traditional and Non-Traditional Foods, Food and Agriculture Organization, Rome, 1981.

11. **Redchinov, V. G.,** *The Improvement of Grain Proteins and Their Evaluations,* Kolos, Moscow, 1978, 19.

12. **Pimental, D., Oltenaeu, P. A., Nesheiar, M. C., Krummel, J., and Allen, M. S.,** The potential for grass-fed livestock: resource constraints, *Science,* 207, 843, 1980.

13. **Vysotsky, V. G. and Tutelyan, V. A.,** *Methodological Problems in Studying the Properties of New Sources of Food Proteins (Review),* All-Union Scientific Research Institute of Medical and Medical Engineering Information, Moscow, 1987.

14. **Parvathy, K. and Subramaniyan, V.,** Effect of leaf protein concentrate on lipid metabolism of albino rats, *Indian Food Packer,* 5, 70, 1984.

15. Lipid Research Clinics Program: The lipid research clinics coronary primary prevention trial results. II. The relationship of reduction in incidence of coronary heart disease to cholesterol lowering, *J. Am. Med. Assoc.,* 251, 3, 365, 1984.

16. **Yatsyshina, T. A., Kalamkarova, O. M., and Ivanshchenko, N. V.,** The influence of food proteins on the level of cholesterol in the blood serum, *Med. Ref. Zh.,* UP I, 35, 1985.

17. **Pokrovsky, A. A.,** *The Pharmacology and Toxicology of Food,* Meditsina, Moscow, 1979.

18. **Roberts, G. R., Mart, E. K., Stalts, V. D., Monro, A. K., Sharbono, S. M., Rodriks, D. V., Poland, H. E., and Doull, D.,** *Safety of Foods,* Agropromizdat, Moscow, 6, 1986.

19. **Rossival, L., Engst, R., and Sokolai, L.,** Foreign substances and food additives in products, *Light and Food Industry,* Moscow, 1982.

20. **Tutelyan, V. A.,** Toxic substances in food and how hazardous they are to human health, *Vopr. Pitan.,* 6, 10, 1983.

21. **Waggle, D. H. and Kolar, C. W.,** Types of soy protein products, in *Soy Protein and Human Nutrition,* Wilcke, H. L., Hopkins, D. T., and Waggle, D. H., Eds., Academic Press, New York, 1979, 19.

22. **Shaternikov, V. A. and Vysotsky, V. G.,** The problem of protein in human nutrition and the directions of its further development, *Vopr. Pitan.,* 5, 24, 1980.

23. Amino Acid Content of Foods and Biological Data on Proteins, FAO Nutritional Studies N24, Food and Agriculture Organization, 1970.

24. **Reili, K.,** *Metallic Contaminants of Food Products,* Agropromizdat, Moscow, 1985.

25. *Codex Maximum Limits for Pesticide Residues,* Codex Alimentarius Commission, Italy, 2nd ed., FAO/WHO, Italy, Vol. XIII, 1986; *Contaminates,* 1st ed., FAO/WHO, Rome, Vol. XVII; *Food Additives,* 1st ed., FAO/WHO, Rome, Vol. XIV, 1983.

Chapter 2

PROTEIN AND AMINO ACID REQUIREMENTS IN RELATION TO DIETARY FOOD PROTEIN NEEDS

Vernon R. Young and Fred H. Steinke

TABLE OF CONTENTS

I. INTRODUCTION

The primary nutritional function of dietary protein is to supply the indispensable (essential) amino acids and a utilizable source of nitrogen required for synthesis of tissue and organ proteins and other nitrogen-containing compounds that are necessary for the normal growth, maintenance, and function of the organism.

In the body, proteins function as organic catalysts (enzymes). They are used for the structural formation of cells, act as antibodies and protein mediators, such as interleukin-1 and tumor necrosis factor, and serve as hormones in the control of cellular metabolism. Inadequate protein or amino acid intake leads to a diminished content of protein in cells and organs and a deterioration in the capacity of cells to carry out their normal function. If the inadequate intake is prolonged, there will be increased morbidity and eventually death. Furthermore, intakes in considerable excess of physiological needs also might be disadvantageous. Thus, an adequate diet, whether consisting of normal foods or based on specially formulated medical products, must contain an appropriate level of protein (nitrogen) and balance of amino acids, one to another, if long-term health is to be assured.

Therefore, in comparing and contrasting the different food proteins and protein sources, it is usual to do this on the basis of their capacity to meet the indispensable amino acid and nitrogen requirements of the host. On the other hand, a more comprehensive evaluation of dietary protein, in relation to the overall nutritional health of the individual and of populations, requires an assessment of the possible effects of various food protein sources on the utilization of, and requirements for, energy-yielding substrates and other individual essential nutrients.[1] In this chapter we will include a discussion of some recent research that, while still somewhat controversial,[2-4] provides a new and, we believe, more rational basis for judging the significance of food proteins in protein and amino acid nutrition. Our major focus will be on the evaluation of protein quality, particularly with reference to humans.

II. OVERVIEW OF APPROACHES FOR EVALUATION OF PROTEIN QUALITY

The multiplicity of nonclinical (Table 1) and clinical methods and procedures (Table 2) that are used to evaluate the nutritional value or quality of dietary protein sources have been the subject of numerous reviews.[5-9] The suitability of a particular method will depend upon the specific use that will be made of the findings, as pointed out earlier by Bender.[5]

In general, the systematic evaluation of dietary sources of protein involves initial chemical testing followed by biological assays. For purposes of evaluating, for example, new or modified plant proteins made available by breeding new varieties of grains or legumes, the first step is usually a chemical analysis for protein content and amino acid profile. Sometimes the amino acid analysis may be for a single critical amino acid, or for a restricted number, such as tryptophan, sulfur amino acids, and lysine. For exploring new and novel sources of protein, the initial approach is to examine the protein concentration and amino acid pattern from which an amino acid score can be derived, followed by *in vitro* tests for digestibility. If these criteria suggest a potentially useful new protein source, a bioassay, both to test for availability of amino acids and as a further test of quality, is then undertaken. These biological tests may reveal a less favorable picture than that indicated by amino acids scoring alone. In this case, the product should be examined for nonavailable amino acids, as, for example, by the use of tests for available lysine, or should be evaluated for possible toxic materials present in the foodstuff. The biological testing should include a measure of digestibility because this can be a significant cause of discrepancy between chemical and biological evaluations of quality.

For a third use, namely the monitoring of variables introduced by food processing, it

TABLE 1
Some Methods for Evaluation of Protein Quality

Approach based on	Examples
Nitrogen retention	N balance index
Weight gain	Protein efficiency ratio
	Relative protein value
Specific biochemical responses	Plasma protein regeneration
	Blood urea concentration
	Cystine dioxgenase activity
	Plasma amino acids
Rapid biological methods with:	Tetrahymena Insects
Other	"Available" lysine enzyme digestion
Amino acid requirements	Amino acid score

Based on Bender, A. E., in *Role of Milk Proteins in Human Nutrition,* Vol. 35, Kaufmann, W., Ed., Kieler Milchwirtschaftliche Forshungsberichte, Gelsenkirchen-Buer, Germany, 1983, 267.

TABLE 2
Methods, in Addition to N Balance and Growth, that Have or May Be Used to Assess Dietary Protein Quality in Human Subjects

Approach	Comments
Nitrogen balance	Commonly used, difficult to measure accurately, and affected by many factors
Growth	Applies to young; ethical constraints
Measurements of blood	
Free amino acids	Potential for estimating limiting amino acid
Urea	Of possible value (children) for a quantitative index,
Albumin	more investigation required
Proteins (enzymes, transport proteins)	
Measurements of urine	
N compounds (urea, ammonia, 3-methyl histidine)	Perhaps useful for qualitative purposes
Body composition	
^{40}K, IVNAA[a] body water (^2H_2O)	Precision and/or availability of technique often a limiting factor

[a]   *In vivo* neutron activation analysis.

From Young, V. R. and Pellett, P. L., *Am. J. Clin. Nutr.,* 42, 1077, 1985.

is common to begin with availability of lysine and of methionine, as these are usually the amino acids most sensitive to processing. For food regulatory purposes, the examination of a protein sample should begin with chemical analysis for nitrogen, amino acids, and, where necessary, for toxins, including those of microbial origin. For routine food regulatory requirements, the amino acid score, corrected for digestibility, is an excellent measure of protein quality as recommended by the Food and Agriculture Organization/World Health Organization (FAO/WHO) 1990.[10]

There have been a number of potential improvements concerning specific aspects of approaches for the assessment of protein quality, including, for example, animal bioassays[11-14] and improved protocols for human studies.[15,16] Nevertheless, we will emphasize here some recent developments relevant to a direct evaluation of the capacity of dietary protein(s) for meeting human protein and amino acids needs. These include an improved understanding of the physiology of protein and amino acid metabolism in human subjects

TABLE 3

Comparison of Whole-Body Protein Synthesis Rates with Dietary Protein Allowances at Different Ages

Age group	Protein synthesis (A)[a]	Protein allowance (B)[a]	Ratio A/B
Infant (premature)	11.3, 14	~3	4.3
Newborn	6.7	1.85	3.6
Child (15 mo.)	6.3	1.3	4.8
Child (2 to 8 years)	3.9	1.1	3.5
Adolescent (~13 years)	~5	1.0	~5
Young adult (~20)	~4.0	0.75	~5.3
Elderly (~70 years)	~3.5	0.75	4.7

[a]   Grams of protein per kilogram per day.

Extended slightly from Young, V. R., Meredith, C., Hoerr, R., Bier, D. M., and Matthews, D. E., in *Substrate and Energy Metabolism in Man*, Garrow, J. S. and Halliday, D., Eds., John Libbey, London, 1985, 119.

and how the body responds to altered protein and amino acid intakes. With this new information we can then consider the amino acid requirements of the human subject, with particular reference to the amino acids supplied by the important food protein sources.

Finally, our particular focus, below, will be on the level and biological availability of amino acids in dietary food proteins because we[17,18] have concluded that the use of dietary amino acid composition data (corrected for digestibility) provides a sound basis for evaluating the quality of dietary protein. This earlier conclusion is supported by the recommendations emerging from a recent FAO/WHO Expert Consultation on Protein Quality Evaluation.[10]

III. SOME METABOLIC ASPECTS OF THE PROTEIN AND AMINO ACID REQUIREMENT

As noted above, the quantitatively important physiological function of dietary protein is to provide substrate necessary for the maintenance of body protein synthesis in the adult, for supporting an acceptable rate of net protein gain in the growing infant and child, for the formation of new tissue in the fetus and supporting tissues of the pregnant female, and for milk production in the lactating woman. Thus, it might be useful to discuss briefly some aspect of whole-body protein synthesis and turnover, and how these change with normal growth and development and during aging in the human subject.

Protein synthesis rates are high in the newborn and, per unit of body weight, these rates decline with progressive growth and development (Table 3). Two points might be made from the data shown here; first, the higher rate of protein synthesis in the young, as compared with that for the adult, is not only related to the *net* protein deposition associated with growth, but there is a high rate of turnover. Thus, protein *synthesis* in the premature infant is about twice as high as in the preschool child, and approximately three or four times as high as in the adult. In parallel with these synthesis rates, the rates of organ and tissue protein *breakdown* are also considerably higher in the infant than in the adult. Second, at all ages, the rates of whole-body protein *synthesis* and breakdown are considerably greater than the safe level of intake of dietary protein estimated to meet the need for maintenance of N balance or for the support of growth. It follows, therefore, that there is an extensive reutilization within the body of the amino acids entering tissue pools during the course of protein breakdown. This recycling of amino acids and the rates of synthesis and breakdown of body proteins change in response to various stimuli, including alterations in the level and adequacy of protein and amino acid intakes.[19-21]

TABLE 4
Obligatory Urinary Nitrogen Losses in Adult Males

Group (location)	Mean age (years)	Mean weight (kg)	Urinary N (mg N kg^{-1} · d^{-1})
U.S.	20	71	38
U.S.	21	74	37
China[a]	23	55	33
India	27	46	38
Nigeria	26	54	34
Japan	—	63	33
Chile	27	63	36

Note: A partial summary from FAO/WHO/UNU (Reference 28) where original literature citations are given for each study.

[a] Taiwan Province.

The importance of protein synthesis as a determinant of the dietary requirement for protein and for indispensable amino acids has been discussed previously.[19,21] It appears necessary to achieve a concentration of free amino acids in tissue pools that is somewhat above that seen with low or grossly inadequate protein and amino acid diets[22] in order to maintain the integrity of the protein synthetic machinery of cells and organs. Because the oxidation of amino acids is sensitive to alterations in their tissue level,[23,24] we have argued that factors regulating the rate of amino acid oxidation are major determinants of the quantitative requirement for specific indispensable amino acids.[25]

The turnover of body proteins and the extensive reutilization of amino acids for the resynthesis of proteins is not completely efficient. Hence, some of the amino acids are lost by oxidative catabolism. This loss includes the metabolism of the carbon skeleton of the indispensable amino acids and the nitrogen of both indispensable and dispensable amino acids. When the dietary protein level is reduced to a submaintenance intake, protein synthesis and breakdown are subsequently reduced, but a more immediate metabolic adjustment is a rise in the efficiency of reutilization of the amino acids liberated during protein breakdown and a fall in the rate at which indispensable amino acids are irreversibly oxidized. This immediate response might be regarded as an *adaptive* mechanism,[26,27] with the somewhat later decline in rates of protein synthesis and breakdown rates representing an *accommodation* to a continued inadequate dietary intake. Consequently, the fall in the irreversible catabolism of indispensable amino acids leads to a lower output of nitrogenous compounds in urine and feces. At essentially protein-free intakes, this output of nitrogen is the so-called "obligatory" nitrogen loss.

As summarized by FAO/WHO/United Nations University (UNU)[28] (Table 4), obligatory N losses in apparently healthy subjects in different countries (possibly with different habitual protein intakes) appear to be remarkably uniform. From these findings it has been concluded[28] that the mechanisms for reducing the minimum requirement for N balance (or for the initial conservation of N) are limited, at least during the early phases of adaptation, to processes responsible for increasing the efficiency of amino acid utilization for protein anabolism, including a reduction in the oxidation of specific indispensable amino acids. The practical implication of these observations is that the dietary requirement for total N (protein), at least in terms of balancing these obligatory losses, is likely to be similar in all healthy populations.

Finally, and to slightly extend the definition made by FAO/WHO/UNU[28] for the requirements for protein (nitrogen), dietary need for the specific, indispensable amino acids in an individual can be defined as the lowest level of intake that will balance the losses of nitrogen and of amino acids (via oxidative catabolism) from the body without major changes in protein turnover and where there is energy balance with a modest level of physical activity. In infants, children, and pregnant and lactating women, the requirement will also include that amount needed for net protein deposition and the secretion of proteins in milk.

The starting point, therefore, for estimating total protein needs is the measurement of the amount of dietary nitrogen needed for zero body N balance in adults during short-term metabolic studies. For determination of the needs for specific indispensable amino acids the approach might be the same or it can be based on a direct estimate of measurement of rates of irreversible oxidation of these amino acids. In the following sections we will further examine briefly the requirements for amino acids, taking into consideration the features of human protein and amino acid metabolism that we have described briefly above.

IV. CURRENT ESTIMATES OF NEEDS FOR PROTEIN AND AMINO ACIDS

A. TOTAL PROTEIN

Most estimates of protein and amino acid requirements have been obtained directly or indirectly from measurements of N balance, and the use and application of the N balance technique for this purpose has been discussed in detail in the 1985 FAO/WHO/UNU[28] report. However, it should be noted that the nitrogen balance technique has serious limitations,[29,30] and it does not provide a totally secure basis for establishing the protein and amino acid needs of human subjects. Estimates of the requirements for protein and indispensable amino acids have been made by various expert groups, including the most recent presented in 1985 by the United Nations.[28]

The recommendations for safe protein intakes include a factor for variation in protein requirements among apparently similar individuals. It has been estimated[28] that the inter-individual variability in protein requirements amounted to a coefficient of variation of 12.5% and, thus, a value of 25% (2SD) above the *mean* minimum physiological requirement of 0.6 g protein per kilogram per day for an adult would meet the needs of all but 2.5% of individuals within the adult population.[28] Hence, the mean minimum requirement should be increased to 0.75 g/kg/d to give a safe protein intake for healthy adults. Apparently, most individuals would require less than this to maintain an adequate state of protein nutrition; it follows that some adult subjects might require as little as 0.45 g high-quality protein per kilogram per day.

The safe protein intakes for infants and young children were arrived at by the FAO/WHO/UNU group as outlined in the 1985 report.[28] These estimates, together with those for adults, and lactating and pregnant women, are summarized in Table 5.

B. SPECIFIC INDISPENSABLE AMINO ACIDS

The requirements for the specific indispensable (essential) amino acids were also assessed by the FAO/WHO/UNU[28] Expert Group. Except for some new data for the preschool-child group, based on studies carried out by Pineda et al.[31] and Torun and co-workers in Guatemala,[32] the values given in the 1985 report[28] are similar to those of the 1971 FAO/WHO committee.[33] Table 6 presents the most recent international estimates for the amino acid requirements in various age groups.

As can be seen from a comparison of the requirements among the various age groups (Table 6), there is a marked decrease in indispensable amino acid needs when expressed per unit of body weight, between infancy and the time adulthood is reached. Additionally, there is a lower requirement for indispensable amino acids, per unit of the total need for protein, in the adult as compared to that for the infant or preschool child. This relatively greater decline in the requirement for indispensable amino acids, compared to the total

TABLE 5
Safe Levels of Protein Intake (grams of protein per kilogram bodyweight per day) as Proposed by FAO/WHO/UNU

Age group	Males	Females
3 to 6 Months	1.85	1.85
6 to 9 Months	1.65	1.65
9 to 12 Months	1.5	1.5
1 to 2 Years	1.2	1.2
2 to 3 Years	1.15	1.15
3 to 5 Years	1.1	1.1
7 to 10 Years	1.0	1.0
10 to 12 Years	1.0	1.0
12 to 14 Years	1.0	0.95
14 to 16 Years	0.94	0.9
16 to 18 Years	0.88	0.8
Adults	0.75	0.75
Pregnancy		+6.0 g[a]
Lactation 0 to 6 months		+17.5 g[a]
Lactation 6 months		+13.0 g[a]

Note: Uncorrected for nutritional value (amino acid scores) of mixed dietary proteins for infants and children and digestibility for all groups.

[a] Total daily addition per subject.

From Energy and Protein Requirements, FAO/WHO/UNU, Tech. Rep. Ser. No. 724, World Health Organization, Geneva, 1985.

TABLE 6
1985 FAO/WHO/UNU Estimates of Amino Acid Requirements at Different Ages (mg/kg per day)

Amino acid	Infants (3 to 4 months)	Children (2 years)	School boys (10 to 12 years)	Adults
Histidine	28	?	?	[8—12]
Isoleucine	70	31	28	10
Leucine	161	73	44	14
Lysine	103	64	44	12
Methionine and cystine	58	27	22	13
Phenylalanine and tyrosine	125	69	22	14
Threonine	87	37	28	7
Tryptophan	17	12.5	3.3	3.5
Valine	93	38	25	10
Total	714	352	216	84
Total per gram protein[a]	434	320	222	111

[a] Total milligrams per gram of crude protein. Taken from Table 38 in Reference 28, and based on all amino acids minus histidine.

From Energy and Protein Requirements, FAO/WHO/UNU, Tech. Rep. Ser. No. 724, World Health Organization, Geneva, 1985.

protein need, might reflect biologically important differences in amino acid metabolism and in the relative efficiency with which amino acids and nitrogen are used by the body, in the younger and older age groups, to maintain an adequate protein nutritional status. However, an equally plausible reason for differences shown in Table 6 for the total amount of indispensable amino acids per unit of protein fed is that they are due to problems arising from the methods used to assess amino acid requirements, especially in adults. Because the accuracy of the estimates shown in Table 6 have important implications for the assessment of protein nutritional quality, we will now consider briefly the currently accepted[28] requirements for adults, since we believe these are the more questionable of the values given in this table.

V. REASSESSMENT OF AMINO ACID REQUIREMENTS FOR ADULTS

The estimates of the requirements for indispensable amino acids, shown in Table 6, have been derived largely from results of nitrogen balance studies in healthy individuals. However, in previous reviews[4,25,34,35,36] we have presented several compelling reasons why these earlier metabolic balance studies, especially those in adults, should be viewed with considerable circumspection, in reference to the use of the N balance data for determinations of the minimum physiological requirements for the indispensable amino acids. Furthermore, from a series of experiments on the kinetics of indispensable amino acid metabolism, in which rates of amino acid oxidation were measured at various amino acid intake levels, we have concluded that the current amino acid requirement estimation (shown in Table 6) for adults are far too low for leucine,[37] valine,[38] lysine,[39] and threonine,[40] and perhaps for all of the other indispensable amino acids except the sulfur amino acids.[41] The foregoing conclusions have important implications for the practical aspects of human protein nutrition and assessment of dietary protein. Hence, in the next section a further rationale will be developed, exploiting a current understanding of the physiology and regulation of amino acid and protein metabolism to support the view that current amino acid requirement values (i.e., Table 6) for adults are far too low.

VI. PREDICTION OF OBLIGATORY RATES OF AMINO ACID OXIDATION

The rates of oxidative loss of the individual indispensable amino acids have not been measured directly under the precise conditions used to measure obligatory urinary nitrogen output (see Reference 33 for details) except perhaps for leucine (see below). However, an estimation of the obligatory oxidative losses of indispensable amino acids might be made on theoretical grounds, if it can be assumed that the oxidation rates of individual amino acids occur in proportion to the pattern of amino acids in mixed body proteins. Thus, as summarized in Table 7, the obligatory oxidation rates of leucine, lysine, and total sulfur amino acids (methionine + cystine) are estimated to be equivalent to approximately 27, 30, and 14 $mg \cdot kg^{-1} \cdot d^{-1}$, respectively. As discussed elsewhere,[4] our earlier published and unpublished data offer support for the predictions of amino acid losses that we have given in Table 7.

From our earlier studies[42] it appears that in the adult the amino acids are recycled with about 90% efficiency when intakes via the diet are distinctly below requirement levels. This is further supported by the fact that at submaintenance protein intakes, whole-body nitrogen flux is approximately 600 mg N $kg^{-1} \cdot d^{-1}$.[43] Hence, if obligatory nitrogen losses are taken to be 54 mg N $kg^{-1} \cdot d^{-1}$,[28] it is evident that the recycling of nitrogen is also about 90%. Therefore, assuming a protein turnover of 3.5 $g \cdot kg^{-1} \cdot d^{-1}$ for a well-nourished adult, the

TABLE 7
Calculation of "Obligatory" Amino Acid Losses (Oxidation) and Estimation of "Minimum" Amino Acid Losses from Protein Turnover

		Amino acid losses predicted by	
		"Protein turnover method"	
Amino Acid	"Obligatory N method"[a]	Flux	Oxidation
Isoleucine	16[b]	168[c]	17[d]
Leucine	27	283	28
Lysine	30	311	31
Methionine and cystine	14	140	14
Phenylalanine and tyrosine	27	280	28
Threonine	15	161	16
Tryptophan	4	42	4
Valine	17	175	18

[a] Total amino acid losses assumed to be equivalent to 50 mg N per kilogram per day and in proportion to the amino acid composition of beef (Reference 28).
[b] All figures expressed as milligrams of amino acid $kg^{-1} \cdot d^{-1}$.
[c] Based on assumption of a protein turnover of 3.5 g protein $kg^{-1} \cdot d^{-1}$, with an amino acid composition equivalent to that of beef proteins (Reference 28).
[d] It is assumed that endogenous amino acids are recycled with 90% efficiency at low amino acid/protein intakes.

Summarized from Young, V. R., Bier, D. M., and Pellett, P. L., *Am. J. Clin. Nutr.*, 50, 80, 1989.

minimum oxidation losses of indispensable amino acids when recycling is 90% can be calculated. These calculations are also summarized in Table 7, and it can be seen that the values are close to those computed from obligatory nitrogen losses (Table 7, column 1). Perhaps the most important conclusion is that the minimum rate of losses of indispensable amino acids is equal to or, in most cases, considerably higher than the estimates of the *upper range of amino acid requirements* as proposed in 1985 by FAO/WHO/UNU.[28]

VII. INTAKES OF AMINO ACIDS TO BALANCE OBLIGATORY AMINO ACID OXIDATION

To maintain body amino acid balance, the obligatory amino acid oxidation rates estimated above (Table 7) must be compensated for by an appropriate dietary supply, just as for the case of balancing obligatory nitrogen losses by an adequate nitrogen intake.[28] The minimum intakes of the individual amino acids required to balance these losses can be calculated readily if the efficiency with which exogenous amino acids are utilized for this purpose is known. Unfortunately, there is a serious lack of direct, experimental data on this aspect of amino acid metabolism and nutrition.

Taking the obligatory amino acid oxidation rates and correcting these for an assumed 70% efficiency of retention of exogenous amino acids, an approximation of the minimum requirement for each of the indispensable amino acids is obtained. The values for the specific amino acids are given in Table 8. For leucine, lysine, and the sulfur-containing amino acids, these approximations amount to 39, 42, and 16 $mg \cdot kg^{-1} \cdot d^{-1}$, respectively. Some evidence[44,45] suggests that lysine may be more effectively conserved, or perhaps utilized, when consumed at low levels than is threonine, for example. Furthermore, the addition of excess levels of non-limiting amino acids causes further changes in body weight gain and compositional parameters when diets limiting in different amino acids, such as threonine or lysine, are

TABLE 8

Prediction of *Minimum Intakes* of Indispensable Amino Acids to Balance Losses Via Irreversible Oxidation

Amino acid	Intake to balance losses predicted by	
	"Obligatory N" method	"Protein turnover" method
Isoleucine	23[a]	24[a]
Leucine	39	39
Lysine	42	43
Methionine and cystine	16	17
Phenylalanine and tyrosine	39	39
Threonine	21	22
Tryptophan	6	5.6
Valine	24	25

[a]   Intakes $(mg \cdot kg^{-1}d^{-1})$ required to balance predicted losses (see Table 7) assuming a 70% efficiency for amino acid retention. For methionine and cystine the amount of methionine converted to cystine at methionine requirement (i.e., 10 mg $kg^{-1}d^{-1}$) was taken to have a 100% efficiency and the remaining 4 mg (for cystine) was assumed to be used with a 70% efficiency. Further details given in Reference 4.

TABLE 9

Comparison of Requirements for Amino Acids, in the Adult, as Estimated by Various Methods

Amino acid	Method[a]		
	N balance (FAO/WHO/UNU)	Prediction[b]	^{13}C-tracer studies
Leucine	14	39	30—40
Lysine	12	62	30
Threonine	7	21	15
Valine	10	24	20
Methionine (without cystine)	13	16	13

[a] $(mg \cdot kg^{-1}d^{-1})$.
[b]   Using the "obligatory N" method (see Table 8).

Based on Young, V. R., Bier, D. M., and Pellett, P. L., *Am. J. Clin. Nutr.*, 50, 80, 1989.

given to growing rats.[46] However, these effects appear to be mediated by changes in food intake[47,48] rather than by changes in the efficiency of use of the limiting amino acid. In order to validate the assumptions used in arriving at predicted intakes to meet minimum physiologic requirements, more human data are needed on the *in vitro* kinetic characteristics of enzymes of amino acid metabolism in different tissues and on levels of amino acids in these tissues to different amino acid intakes, and how these various indices change in relation to the *in vivo* metabolic fate of amino acids. We must emphasize therefore that we offer the predictions above as a hypothesis at this point in time. Although we believe this to be a reasonable one, it requires additional experimental support, some of which is discussed below.

VIII. COMPARISON OF CALCULATED MINIMUM REQUIREMENTS WITH THOSE OBTAINED FROM ^{13}C-TRACER STUDIES

The approximations for adult amino acid requirements (Table 9), obtained from the

TABLE 10
Proposed, Revised Estimates of the Mean
Requirements for Amino Acids in Adults and Their
Relationship to the Requirement for Protein

	Mean requirement (mg)	
Amino acid	Per kg body wt	Per gram of protein
Isoleucine	23	38
Leucine	40	66
Lysine	30	50
Total SAA	13	22
Total AAA	39	65
Threonine	15	25
Tryptophan	6	10
Valine	20	33

Based on data presented in Young, V. R., Bier, D. M., and Pellett, P. L., *Am. J. Clin. Nutr.*, 50, 80, 1989.

predictions of obligatory amino acid oxidation rates (factorial or obligatory N loss method) and from the estimates based on considerations of protein turnover (protein turnover method), can be compared with those derived from our studies involving the use of ^{13}C-amino acid tracers. Table 9 gives this comparison for the five amino acids that have been studied in our laboratories to date using this approach. In view of the difficulties encountered in the design, conduct, and interpretation of amino acid kinetic studies, and the reasonable, but not yet validated assumptions discussed above concerning amino acid recycling efficiency and retention, there is a remarkably good agreement between the estimated requirement levels as judged using these three new approaches. Furthermore, given the agreement between the requirement estimations based on the obligatory N loss, and ^{13}C-kinetic methods, this further underscores the much lower requirement figures previously accepted by FAO/WHO/UNU.[28] The discrepancy shown here (Table 9) between the FAO (N balance) values and our proposed estimates must be viewed with considerable concern. Indeed, from this present analysis, the case for the inadequacy of current national and international amino acid requirement figures appears to us to be highly convincing, and there are no published and convincing arguments to the contrary, in our opinion.

Clearly, in view of the potential importance of these new amino acid requirement data for evaluation of protein quality in human nutrition, conformation and extension of our studies is highly desirable. Few laboratories are engaged in work of this specific nature, and so these additional data are likely to be rather slow in coming. However, a recent tracer study by Zello et al.[49] on the metabolism of phenylalanine in young adult men adds support to the conclusions we have drawn about the significant underestimation of requirements made by the international expert group.[28]

IX. SUMMARY OF NEW PREDICTIONS FOR AMINO ACID REQUIREMENTS AND RELATION TO PROTEIN REQUIREMENTS

A summary of our revised estimates, expressed per kilogram of body weight for the amino acid requirements of the adult, are shown in Table 10. To establish the relationship between the proposed, revised estimates of adult amino acid requirements and the requirement for protein, we can arrive at a new adult amino acid requirement (amino acid scoring) pattern. Thus, the relationship between the approximate mean amino acid requirements, as estimated

TABLE 11
Some National and International Recommended Amino Acid Scoring Patterns

Amino acid	1957 FAO[50]	1973 FAO/WHO[33]	1974 NAS-NRC[52]	1985 FAO/WHO/UNU[28]			
				Infant (<1 yr)	Preschool child (2 to 5 yrs)	School-age child (6 to 12 yrs)	Adult (13 yrs+)
Histidine	—	14	17	26	19	19	16
Isoleucine	42	40	42	46	28	28	13
Leucine	48	70	70	93	66	44	19
Lysine	42	55	51	66	58	44	16
Total SAA	42	35	26	42	25	22	17
Total aromatic	56	60	73	72	63	22	19
Threonine	28	40	35	43	34	28	9
Tryptophan	14	10	11	17	11	9	5
Valine	41	50	48	55	35	25	13
Total (w/o histidine)	314	360	356	434	320	222	111

Note: Figures are expressed as milligrams of amino acid per gram of protein (N × 6.25).

by the different approaches (predicted and ^{13}C-Tracer) above, and the mean requirement of value 0.6 g of high-quality, highly digestible protein $kg^{-1} \cdot d^{-1}$ (see Reference 28) can be determined. This has also been done as shown in Table 10.

X. IMPLICATIONS FOR EVALUATION OF DIETARY PROTEIN

The proposed new estimates of amino acid requirements for the adult (Table 10) increases the levels, on average, by a factor of 2.5, relative to those proposed in 1985 by FAO/WHO/UNU.[28] At first sight, this might appear to recreate a world food "protein problem". The practical implications, however, while possibly significant for dietaries and food and agricultural policies in some developing countries, are less for diets of some other developing countries and the developed regions where diets are generally rich in foods of animal origin. In the following paragraphs, we will develop and then explore briefly the application of a new amino acid requirement, or *scoring*, pattern for purposes of assessment of food proteins in diets in both the developed and developing regions of the world.

XI. PREVIOUS AMINO ACID SCORING SYSTEMS

Before developing a new amino acid scoring pattern, it is worth noting that various amino acid scoring systems have been recommended previously by various national and international groups.[28,33,50-52] They have been developed for purposes for predicting the capacity of food protein sources to meet human physiologic needs for nitrogen and indispensable amino acids, and a number of the earlier amino acid scoring patterns are shown in Table 11.

The significant changes that have occurred in the recommendations over the years are gathered from this summary (Table 11). Although the 1985 FAO/WHO/UNU[28] group was the first to explicitly recommend lower values for the adult, the earlier 1973 FAO/WHO[33] group had fully recognized that the adult requirement values (expressed per unit of protein) were much lower, but nevertheless based its scoring pattern on data for infants and young children.

It has been 33 years now since the first FAO committee[50] considered protein or amino acid scoring. Provided that an adjustment is made for the digestibility of ingested proteins,[53] the statement made by this international group[50] in 1957 concerning the role of scoring in relation to an amino acid requirement pattern remains valid. It is as follows:

The concept of a desirable pattern of essential amino acid has one great advantage. By comparison with such a pattern, data on the amino acid content of food combinations can be appraised in a wide range of situations in terms of possible defects of the diet and of methods of improving it. Comparisons, no doubt rough and approximate, can be made directly by the use of tables showing the contents of foods in essential amino acids, regardless of the proportions in which individual foods are included in the diet. On the other hand, even when the biological value of each dietary component is known, a deduction cannot be made about the biological value of the diet as a whole.

As discussed elsewhere in more detail,[54] use of amino acid scoring systems has progressed over the years with the general expectation that not only should an amino acid score be able to predict the potential nutritional value of a food or diet for humans, but that such a score should (with or without digestibility consideration) also correlate directly with the results of animal assays, such as net protein utilization. Whether such correlations with animal assays are a valid or even desirable attribute of amino acid scoring systems intended for application in human nutrition remains questionable. We[54] have stated that scoring systems developed to predict the nutritional value of protein sources for humans should not necessarily be expected to agree with values obtained with growing rats. We now contend that the appropriate standard for dietary assessment is the human amino acid requirement pattern and that for animal bioassays to be useful, they should be designed to give predictions in line with those based on human amino acid requirements rather than the reverse.

XII. AN AMINO ACID SCORING PATTERN FOR APPLICATION IN YOUNG CHILDREN AND ADULTS

Thus, accepting the validity and usefulness of the amino acid scoring approach, the newly proposed estimates of the amino acid requirements in adults (Table 10) provide a basis for making a further refinement in establishing a more satisfactory approach for evaluating protein quality with the aid of an amino acid scoring pattern.

Based on our tracer studies and the predictions made above, together with a further consideration of the sulfur amino acid requirements in the young and the nutritional quality of isolated soy proteins in the adult, we can propose a recommended amino acid pattern for the adult. This is given in Table 12. It can be seen that this new amino acid scoring pattern is essentially the same as that for the 1985 FAO/WHO/UNU 2- to 5-year-old pattern (Table 12), with the significant exceptions that threonine is lower in the proposed adult pattern and lysine is somewhat higher in the 1985 FAO/WHO/UNU[28] preschool-child pattern. On this basis, and in relation to the practical problems of the evaluation of dietary protein and for purposes of adjusting safe practical allowances for protein intakes for usual diets,[28] it appears that a single amino acid scoring pattern would be adequate for the entire age range, covering preschool children through adults. Therefore, the 2- to 5-year-old child pattern shown in Table 12 should be used for this purpose, and this is consistent with the recommendation made recently by FAO/WHO/UNU.[10]

XIII. APPLICATION OF THE PRESCHOOL-CHILD AMINO ACID SCORING PATTERN

As indicated at the beginning of this chapter, there are multiple methods available for assessment of protein nutritional quality. For routine food regulatory purposes, we suggest that an amino acid scoring procedure would be appropriate. Now that we have arrived, above, at a single amino acid scoring pattern, it is worth returning to the recent Report of a Joint FAO/WHO Expert Consultation[10] which reviewed the question of the appropriate methods for measuring the quality of food proteins for the population. The Consultation concluded that the most appropriate method available was the protein digestibility-corrected amino acid score (PD-CAS) method. This method was then recommended to be used as the

TABLE 12
Tentative Amino Acid Requirement Estimates for the Adult and Corresponding Requirement Pattern for the Preschool Child

| | Adult | | |
| | Tentative requirement[a] (mg/kg/d) | Amino acid pattern[b] (mg/g protein) | 1985 FAO/WHO preschool pattern[c] (mg/g protein) |
Amino acid			
Isoleucine	23	35	28
Leucine	40	65	66
Lysine	30	50	58
Total SAA[d]	13	25	25
Total AAA[e]	39	65	63
Threonine	15	25	34
Tryptophan	6	10	11
Valine	20	35	35

[a] From Table 10.
[b] Values rounded to nearest 5.
[c] FAO/WHO/UNU, Reference 28.
[d] Sulfur amino acids.
[e] Aromatic amino acids.

method of choice internationally. The amino acid score, corrected for digestibility, used the amino acid for the 2- to 5-year-old child as proposed in 1985 by FAO/WHO/UNU[28] as the reference pattern in this procedure.

In addition to defining the amino acid reference pattern for use in the PD-CAS method, the FAO/WHO Consultation[10] also established the procedures for measuring amino acids and digestibility. In Table 13, a worked example for estimating the nutritional quality of a mixture of protein sources is presented. As can be seen, the indispensable amino acid with the lowest score becomes the PD-CAS for the protein. If the score of all amino acids is greater than 1, then the PD-CAS is reduced to 1.0.

The PD-CAS offers considerable benefits over that of animal bioassays which traditionally have been used to assess protein quality of food proteins. An important benefit is that PD-CAS uses human amino acid requirements as the basis of evaluation which ensures that appropriate levels of indispensable amino acids are provided in the diet. Also, use of the amino acid scoring procedure will facilitate an evaluation of blending of foods to optimize utilization and meeting protein and amino acid needs.

The development of an internationally derived procedure for evaluating protein quality using the amino acid scoring concept is a step that has long been required. This PD-CAS procedure can be modified in specific terms as new knowledge about specific amino acid requirements emerges and as the determination of availability of amino acids is improved upon and the phenomenon better understood. For the present, the PD-CAS proposed by FAO/WHO[10] should be very useful for evaluating the nutritional quality of human food protein sources.

XIV. SOME OTHER HEALTH-RELATED ASPECTS OF DIFFERENT PROTEIN FOOD SOURCES

A. PROTEIN-ENERGY RELATIONS

As also pointed out earlier, a major nutritional difference between foods of animal origin and those of plant origin could be attributed to their primary content of indispensable amino

TABLE 13
A Worked Example for Estimation of the PD-CAS for a Mixture of Wheat, Chickpea, and Milk Powder

	Analytical data								Quantities in mixture			
	Wt. (g)	Protein (g/100 g)	Lys	SAA	Thr	Trp	Digestibility factor	Protein (g) $\frac{A \times B}{100} = P$	Lys	TSAA (mg)	Thr	Trp
				(mg/g) protein								
	A	B	C	D	E	F	G		PXC	PXD	PXE	PXF
Wheat	350	13	25	35	30	11	0.85	45.5	1138	1593	1365	501
Chickpea	150	22	70	25	42	13	0.80	33.0	1310	825	1386	429
Milk powder	50	34	80	30	37	12	0.95	17.0	1360	510	629	204
Totals								95.5	4808	2928	3380	1134
Amino acids mg/g protein[a]									50	31	35	12
Reference scoring pattern[b]			58	25	34	11						
Amino acid score for mixture[c]									0.86	1.24	1.03	1.09
Weighted average protein digestibility[d]							0.85					
Score adjusted for digestibility (0.85 × 0.86)									0.73 (or 73%)			

Note: Taken from Table 10 in FAO/WHO, Reference 10.

a   Total for each amino acid/total protein.
b   mg/g protein.
c   Amino acids per gram of protein divided by reference pattern.
d   Sum of (protein × factor [PXG] divided by protein total.

TABLE 14
Composition of Selected Food Protein Sources: Macroconstituents (g/100 g)

Food	Protein	Carbohydrate	Fat	Dietary fiber	Water	Energy (Kcal)
Cereal						
Oatmeal (raw)	12.4	72.8	8.7	7.0	8.9	401
Flour (white, 72%)	11.3	73.3	1.2	3.0	14.5	337
Milk and dairy products						
Milk (fresh, whole)	3.3	4.7	3.8	—	87.6	65
Eggs (boiled)	12.3	tr	10.9	—	74.8	147
Meat and meat products						
Beef (minced, raw)	18.8	—	16.2		64.5	221
Frankfurter	9.5	3.0	25.0		59.5	274
Fish						
Lemon sole (raw)	17.1	—	1.4	—	81.2	81
Vegetables						
Beans (baked, tomato sauce)	5.1	10.3	0.5	7.3	73.6	64
Nuts						
Peanuts (fresh)	24.3	8.6	49.0	8.1	4.5	570

Summarized from Paul, A. A. and Southgate, D. A. T., McCance and Widdowson's "The Composition of Foods", 4th ed., Her Majesty's Stationery Office, London, 1978.

TABLE 15
Energy Density and Protein Content of Selected Foods

Food	Energy value (kcal/100 g)	Protein (g per) 100 g	Protein (g per) 100 kcal
Boiled rice	123	2.2	1.8
Potatoes (new, boiled)	76	1.6	2.1
Bread (white)	233	7.8	3.3
Milk (whole, fresh cow)	65	3.3	5.7
Beef (lean)	123	20.3	16.5

Extracted and calculated from Paul, A. A. and Southgate, D. A. T., McCance and Widdowson's "The Composition of Foods", 4th ed., Her Majesty's Stationery Office, London, 1978.

acids and the digestibility of the proteins. This is an appropriate comparison when the nutritional value of proteins is viewed in the context of human nitrogen and indispensable amino acid needs. Protein foods, however, are nutritionally more significant and chemically more complex than can be judged solely from the levels of available amino acids they contain.

Thus, the composition of a variety of food protein sources is shown in Table 14. This summary indicates that there are substantial differences not only in the concentration of protein per unit weight of food, but also in the concentration of other macroconstituents of these foods, including carbohydrate, fat, and dietary fiber. These compositional differences have potentially important consequences for the maintenance of nutritional health, particularly with respect to individuals whose nutrient requirements are relatively high. Hence, as might be gathered from the data in Table 15, if protein and energy needs are supplied in major part from a rice-based diet, the energy density of the diet, as traditionally prepared, would be relatively low. Indeed, a number of investigators[56,57] have pointed out that the bulkiness of traditional diets makes it difficult to meet energy requirements for growing infants and children. Although the quality of rice protein is good, and in theory capable of

TABLE 16

Dietary Fiber, Phosphorus, Phytic Acid, and Some Essential Minerals in Selected Plant Food Protein Sources

| | | Phosphorus | | | | |
Food	Dietary fiber (mg/100 g)	Total (mg/100 g)	Percent as phytic acid	Iron	Zinc	Calcium
Oatmeal (raw)	7.0	380	70	4.1	(3.0)	55
Flour (white, 72%)	3.0	130	30	1.5	0.9	15
Rice (polished raw)	2.4	100	61	0.5	1.3	4
Peanuts (raw)	8.1	370	57	2.0	3.0	61
Beans (broad, boiled)	4.2	99	5	1.0	—	21

Summarized from Paul, A. A. and Southgate, D. A. T., McCance and Widdowson's "The Composition of Foods," 4th ed., Her Majesty's Stationery Office, London, 1978.

meeting the requirement for essential amino acids when consumed as the sole protein source, the overall composition of this food and the way in which it is prepared limit its potential for meeting the energy requirement of the young. The interrelationships between the protein content, energy density, and bulkiness of plant protein sources are therefore important additional determinants of the nutritional quality of the predominantly vegetable-based diets especially for many developing areas of the world.

B. IMPACT ON MINERAL NUTRITION

There are also marked differences in the content, and potentially the availability, of essential minerals and vitamins among the important food protein sources. In view of the widespread occurrence of iron deficiency anemia in many developing countries and the increasing trend toward a greater dependence on plant protein foods, particularly soy foods, by populations in the technically developed regions, it is relevant also to consider briefly the relationships between protein intakes and food sources and human mineral nutrition. Table 16 summarizes the content of iron, zinc, and calcium in selected plant food sources and in addition gives the content of dietary fiber and phytic acid phosphorus in these foods. Clearly, the concentration of individual minerals varies greatly both in absolute amount and in relation to other minerals and dietary constituents that may affect their utilization.

With regard to iron, the extent to which this element is available in foods depends not only on the amount of iron supplied, but also on the chemical form of the iron and the composition of the meal in which it is consumed.[58] Furthermore, there are marked differences in the form of iron in animal and plant protein foods, with the former providing a substantial proportion of the total iron as heme iron. The heme iron is usually absorbed with relatively high efficiency and is largely unaffected by dietary factors. On the other hand, plant food proteins contain nonheme iron, the availability of which is less than that of heme iron and its availability is modified markedly by various dietary factors, such as those listed in Table 17. Clearly then, a diet based on plant protein foods may not easily meet the requirements for iron, especially in premenopausal women. In this context, animal flesh helps to meet iron requirements more easily than do plant protein foods that might otherwise supply equivalent levels of utilizable protein.

The question of the nutritional value of food proteins in relation to mineral nutriture extends beyond dietary iron, and it is worthwhile expanding briefly here. First, a large number of factors can affect the extent to which a mineral element in a food is available to meet the physiologic needs of the host or the impact of a specific food protein on the mineral status of the host.[1] Among these factors is food processing; this may contribute to either an

TABLE 17
Factors Influencing Absorption of
Nonheme Iron

Enhance	Depress
Ascorbic acid	Tannic acid
Meat factor	Phosvitin (egg yolk)
	Egg albumin
	Phytate
Host iron depletion	Ca and P salts
	Antacids

From Young, V. R., in *Nutrition in the 1980's: Constraints on Our Knowledge,* Selvey, N. and White, P. L., Eds., Alan R. Liss, New York, 1981, 189.

enhanced or a reduced availability of a particular mineral in a food protein source, depending upon the process of the primary food material in question. Therefore, generalizations about the bioavailability of minerals in different protein foods may be misleading, and it cannot be concluded that mineral bioavailability is necessarily the same in all cereals or cereal products on the one hand, or in all soy-based foods on the other.

Second, the availability of dietary zinc may be influenced by the level of phytic acid (myo-inositol-1,2,3,4,5,6-hexakis[dihydrogen phosphate]). The concentration of this compound varies among different plant protein sources and also in relation to the zinc contained in these foods (Table 16). The possible implications for human mineral nutrition of the phytic acid content of plant protein foods has attracted considerable investigation.[59,60] Results obtained in animal studies have firmly established the inhibitory influence of this plant constituent on the availability of several dietary minerals, especially calcium and zinc. The practical consequences for human mineral nutrition arising from phytic acid ingestion through normal diets, however, are difficult to judge. This will depend upon the specific element, its level in the diet, and the presence or absence of other interacting factors.

In order to provide a more complete statement of the interrelationships between protein intake and source and mineral nutrition, the importance of obtaining additional human metabolic data cannot be overemphasized.

The interrelationships between body calcium homeostasis and dietary protein also have been investigated in rat and human studies.[61-65] From metabolic balance studies, it appears that high-protein diets may cause a deterioration in body calcium balance. The major effect of a high-protein diet appears to be located at the level of kidney, the mechanism associated with a reduced reabsorption of filtered calcium. Although the public health significance of these observations remains uncertain, there has been the suggestion that high-protein diets may contribute to the etiology of osteoporosis in the adult population. However, it seems that there is little reason to conclude, at least from the available evidence, that dietary protein is of major importance in the incidence of osteoporosis; other factors, such as calcium intake, estrogenic status, and perhaps exercise are likely to be more significant. Nevertheless, this is a subject that deserves more careful scrutiny.

C. VITAMINS

There are a number of examples of metabolic associations between food protein sources and vitamin nutriture in human subjects.[1] It is recognized that diets low in animal foods may not provide sufficient vitamins B_{12} and D to meet physiologic needs and that the bound form of niacin present in some protein sources, such as corn, may be unavailable to the host unless there has been appropriate pretreatment of the food. Although the practical significance of some of these metabolic interactions remains unclear, they require mention because they also indicate the importance of studying interactions among nutrients as a sound basis for evaluating fully the role of food proteins in human health.

D. PROTEINS AND LIPID METABOLISM

Another aspect of the role of food proteins in human health concerns their possible effects on lipid metabolism in human subjects. Thus, various investigators[66-69] have shown that there are striking differences among animal and vegetable proteins in their effects on plasma cholesterol levels in some experimental animal models. In rabbits, the effect of different proteins on serum cholesterol levels is seen in the absence of dietary cholesterol. Findings in humans suggest that the type of dietary protein has a marked effect only in patients with hypercholesterolemia. The effect is quite variable among different animal species,[70] and relatively small or absent in normocholesterolemic subjects (reviewed by Terpstra et al.[71]).

The cholesterol lowering effect of dietary proteins, such as soy protein, in animals appears to be related to changes in cholesterol absorption, synthesis, and excretion.[71] The fall in cholesterol concentrations is accompanied by reduced cholesterol absorption and increased cholesterol turnover. Furthermore, the changes in cholesterol levels due to differences in dietary protein can be modulated by the type of dietary fat and fiber and the age of the experimental animal.[72] In hypercholesterolemic men, soy protein increases the fractional turnover rate of VLDL apolipoprotein B.[73]

The significance of these various findings for the health of normal populations is difficult to judge precisely, but again it seems that the source and level of dietary protein per se plays only a minor role, if any, in determining blood cholesterol.[71] Other dietary factors, such as the amount and chemical form of fat, as well as genetic factors, have a far more profound and important influence. Nevertheless, there does appear to be a useful role for dietary protein, particularly of soy protein, in the treatment and management of patients with major disturbances in their blood lipid and cholesterol levels.

E. PROTEINS AND RELATION TO CANCERS

Finally and briefly, there is the question of the role, if any, of dietary protein in the etiology of cancers at various organ sites. Specifically, it is reasonable to ask whether or not usual protein intakes, discussed above, that for many people exceed estimated minimum physiological needs by more than one- or twofold, increase the risk of cancers. To summarize from two comprehensive surveys of this problem,[74,75] the level of protein intake appears to affect the incidence of spontaneous and chemically induced tumors in experimental animals. However, the relevance of these findings for human nutrition cannot be judged adequately. Furthermore, while some epidemiological studies have suggested possible associations between high protein intakes and cancers, the associations are weak and the available literature is still quite limited. Indeed, because of the high correlations between protein and other food constituents, such as fat, it seems possible that high protein intake *may* be associated with increased risk of cancer of certain sites, but it is not yet possible to draw any firm conclusion about an independent effect of protein. The National Research Council Committee[75] recommends maintaining protein intake at moderate levels, or levels lower than twice the recommended dietary allowances (i.e., less than about 1.6 g/kg body weight for adults).

XV. SUMMARY AND CONCLUSIONS

The role of food proteins in meeting the nitrogen and amino acid needs of human subjects has been considered in this chapter under a number of different subtopics. First, we presented a brief survey of the approaches taken to evaluate the nutritional quality of food protein sources. Then we considered some metabolic aspects of the protein and amino acid requirement as a basis for a further and critical review of current estimates of human protein and amino acid needs. New requirement estimates for adults were discussed, and the development of an amino acid scoring procedure for application in the evaluation of protein quality was reviewed. A recently proposed procedure, called the PD-CAS method by FAO/WHO,[10] was presented in light of new knowledge about human amino acid requirements. This method has been recommended for international use and in reference to routine regulatory procedures for determination and control of the protein quality of usual and processed foods. It was also pointed out that a consideration of different food proteins in relation to amino acid and nitrogen requirements constitutes an important, but only partial basis for evaluating the role of dietary protein and of various food protein sources in human nutrition and health. Different food protein sources influence in various ways the utilization of and possible requirements for other nutrients. This complicates the determination of requirements and the setting of rational and safe dietary allowances, as well as of dietary guidelines for individuals and population groups.

A complete and satisfying assessment of the role and impact of food proteins and their interrelationships with, and interactions among, other foods and essential nutrients, particularly with respect to the nutritional metabolic status of free-living individuals and their long-term health, presents an exciting research challenge to nutrition and other health professionals.

REFERENCES

1. **Young, V. R.,** Food protein sources; implications for nutrient requirements, in *Nutrition in the 1980's: Constraints on Our Knowledge,* Selvey, N. and White, P. L., Eds., Alan R. Liss, New York, 1981, 189.
2. **Millward, D. J. and Rivers, J. P. W.,** The nutritional role of indispensable amino acids and the metabolic basis for their requirements, *Eur. J. Clin. Nutr.,* 42, 367, 1988.
3. **Millward, D. J., Jackson, A. A., Price, G., and Rivers, J. P. W.,** Human amino acid and protein requirements: current dilemmas and uncertainties, *Nutr. Res. Rev.,* 2, 109, 1989.
4. **Young, V. R., Bier, D. M., and Pellett, P. L.,** A theoretical basis for increasing current estimates of the amino acid requirements in adult men with experimental support, *Am. J. Clin. Nutr.,* 50, 80, 1989.
5. **Bender, A. E.,** Protein evaluation, in *Role of Milk Proteins in Human Nutrition,* Vol. 35, Kaufmann, W., Ed., Kieler Milchwirtschaftliche Forshungsberichte, 1983, 267.
6. **Young, V. R. and Pellett, P. L.,** Wheat proteins in relation to protein requirements and availability of amino acids, *Am. J. Clin. Nutr.,* 42, 1077, 1985.
7. **Nutritional Value of Protein Foods, Pellett, P. L. and Young, V. R., Eds.,** United National University, World Hunger Program; Food and Nutrition Bulletin, Suppl. 4, United Nations University, Tokyo, 1980, 154.
8. *Proteins in Human Nutrition,* Porter, J. W. G. and Rolls, B. A., Eds., Academic Press, New York, 1973, 560.
9. *Protein Quality in Humans: Assessment and In Vitro Estimation,* Bodwell, C. E., Adkins, J. S., and Hopkins, D. T., Eds., AVI Publishing, Westport, CT, 1981, 435.
10. Protein Quality Evaluation, Report of a Joint FAO/WHO Expert Consultation, Food and Agriculture Organization, Rome, 1990.
11. **Finke, M. D., DeFoliart, G. R., and Benevenga, N. J.,** Use of simultaneous curve fitting and a four-parameter logistic model to evaluate the nutritional quality of protein sources at growth rates of rats from maintenance to maximum gain, *J. Nutr.,* 117, 1681, 1987.
12. **Phillips, R. D.,** Linear and non-linear models for measuring protein nutritional quality, *J. Nutr.,* 111, 1058, 1981.
13. **Phillips, R. D.,** Modification of the saturation kinetics model to produce a more versatile protein quality assay, *J. Nutr.,* 112, 468, 1982.
14. **Mercer, L. P.,** The quantitative nutrient-response relationship, *J. Nutr.,* 112, 560, 1982.

15. **Young, V. R., Rand, W. M., and Scrimshaw, N. S.,** Measuring protein quality in humans: a review and proposed method, *Cereal Chem.,* 54, 929, 1988.

16. **Fomon, S. J., Ziegler, E., Nelson, S. E., and Edwards, B. B.,** Requirement of sulfur-containing amino acids in infancy, *J. Nutr.,* 116, 1405, 1986.

17. **Pellett, P. L. and Young, V. R.,** Background paper 4: evaluation of the use of amino acid composition data in assessing the protein quality of meat and poultry products, *Am. J. Clin. Nutr.,* 40, 718, 1984.

18. **Young, V. R. and Pellett, P. L.,** Background paper 5: amino acid composition in relation to protein nutritional quality of meat and poultry products, *Am. J. Clin. Nutr.,* 40, 737, 1984.

19. **Young, V. R., Moldawer, L. L., Hoerr, R., and Bier, D. M.,** Mechanisms of adaptation to protein malnutrition, in *Nutritional Adaptation in Man,* Blaxter, K. L. and Waterlow, J. C., Eds., John Libbey, London, 1985, 189.

20. **Waterlow, J. C., Garlick, P. J., and Millward, D. J.,** *Protein Turnover in Mammalian Tissues and in the Whole Body,* North-Holland, Amsterdam, 1978, 804.

21. **Young, V. R., Fukagawa, N. K., Bier, D. M., and Matthews, D.,** Some aspects of *in vitro* human protein and amino acid metabolism, with particular reference to nutritional modulation, in *Verhandlungen der deutschen Gesellschaft fur Innere Medizin,* Vol. 92, J. F. Bergmann, Verlag, Munich, 1986, 639.

22. **Munro, H. N.,** Free amino acid pools and their role in regulation, in *Mammalian Protein Metabolism,* Vol. IV, Munro, H. N., Ed., 1970, 229.

23. **Kang-Lee, T. A. and Harper, A. E.,** Effect of histidine intake and hepatic histidase activity on the metabolism of histidine *in vivo, J. Nutr.,* 107, 1427, 1977.

24. **Harper, A. E. and Benjamin, A.,** Relationship between intake and rate of oxidation of leucine and α-ketoisocaproate *in vivo* in the rat, *J. Nutr.,* 114, 57, 1984.

25. **Young, V. R., Meredith, C., Hoerr, R., Bier, D. M., and Matthews, D. E.,** Amino and kinetics in relation to protein and amino acid requirements: the primary importance of amino acid oxidation, in *Substrate and Energy Metabolism in Man,* Garrow, J. S. and Halliday, D., Eds., John Libbey, London, 1985, 119.

26. **Young, V. R., Gucalp, C., Rand, W. M., Matthews, D. E., and Bier, D. M.,** Leucine kinetics during three weeks at submaintenance-to-maintenance intakes of leucine in men: adaptation and accommodation, *Hum. Nutr. Clin. Nutr.,* 41C, 1, 1987.

27. **Waterlow, J. C.,** What do we mean by adaptation?, in *Nutritional Adaptation in Man,* Blaxter, K. L. and Waterlow, J. C., Eds., John Libbey, London, 1985, 10.

28. Energy and Protein Requirements, FAO/WHO/UNU, Tech. Rep. Ser. No. 724, World Health Organization, Geneva, 1985.

29. **Hegsted, D. M.,** Balance studies, *J. Nutr.,* 106, 307, 1976.

30. **Young, V. R.,** Nutritional balance studies: indicators of human requirements or of adaptive mechanisms?, *J. Nutr.,* 116, 700, 1986.

31. **Pineda, O., Torun, B., Viteri, F. E., and Arroyave, G.,** Protein quality in relation to estimates of essential amino acid requirements, in *Protein Quality in Humans: Assessment and In Vitro Estimation,* Bodwell, C. E., Adkins, J. S., and Hopkins, D. T., Eds., AVI Publishing, Westport, CT, 1981, 29.

32. **Torun, B., Pineda, O., Viteri, F. E., and Arroyave, G.,** Use of amino acid composition data to predict nutritive value for children with specific reference to new estimates of this essential amino acid requirements, in *Protein Quality in Humans: Assessment and In Vitro Estimation,* Bodwell, C. E., Adkins, J. S., and Hopkins, D. T., Eds., AVI Publishing, Westport, CT, 1981, 374.

33. Energy and Protein Requirements, FAO/WHO Nutr. Meetings, Rep. Ser. No. 52, Food and Agriculture Organization, Rome, 1973.

34. **Young, V. R., Meguid, M., Meredith, C., Matthews, D. E., and Bier, D. M.,** Recent developments in knowledge of human amino acid requirements, in *Nitrogen Metabolism in Man,* Waterlow, J. C. and Stephen, J. M. L., Eds., Applied Science Publishers, London, 1981, 133.

35. **Meredith, C., Bier, D. M., Meguid, M. M., Matthews, D. E., Wen, Z., and Young, V. R.,** Whole body amino acid turnover with ^{13}C-tracers: a new approach for estimation of human acid requirements, in *Clinical Nutrition '81,* Wesdorp, R. I. C. and Soeters, P. B., Eds., Churchill Livingstone, London, 1982, 42.

36. **Young, V. R. and Marchini, J. S.,** Mechanism and nutritional significance of metabolic responses to altered intakes of protein and amino acids, with reference to nutritional adaptation in humans, *Am. J. Clin. Nutr.,* 51, 270, 1990.

37. **Mequid, M. M., Matthews, D. E., Bier, D. M., Meredith, C. N., Soeldner, J. S., and Young, V. R.,** Leucine kinetics at graded leucine intakes in young men, *Am. J. Clin. Nutr.,* 43, 370, 1986.

38. **Mequid, M. M., Matthews, D. E., Bier, D. M., Meredith, C. N., and Young, V. R.,** Valine kinetics at graded valine intakes in young men, *Am. J. Clin. Nutr.,* 43, 781, 1986.

39. **Meredith, C. N., Wen, Z. W., Bier, D. M., Matthews, D. E., and Young, V. R.,** Lysine kinetics at graded valine intakes in young men, *Am. J. Clin. Nutr.,* 43, 787, 1986.

40. **Zhao, X. L., Wen, Z. W., Meredith, C. N., Matthews, D. E., Bier, D. M., and Young, V. R.,** Threonine kinetics graded threonine intakes in young men, *Am. J. Clin. Nutr.,* 43, 795, 1986.

41. **Young, V. R., Wagner, D. A., Bevine, R., and Storck, K. J.**, Methionine kinetics and balance at the 1985 FAO/WHO/UNU intake requirement level in young men studied with [^{2}H$_3$-methyl-1-^{13}C] methionine as a tracer, *Am. J. Clin. Nutr.*, 54, 377, 1991.

42. **Motil, K. J., Matthews, D. E., Bier, D. M., Burke, J. F., Munro, H. N., and Young, V. R.**, Whole-body leucine and lysine metabolism: response to dietary protein intake in young men, *Am. J. Physiol.*, 240, E712, 1981.

43. **Steffee, W. P., Goldsmith, R. S., Pencharz, P. B., Scrimshaw, N. S., and Young, V. R.**, Dietary protein intake and dynamic aspects of whole body nitrogen metabolism in adult humans, *Metabolism*, 25, 281, 1976.

44. **Yoshida, A. and Ashida, K.**, Pattern of essential amino acid requirement for growing rats fed on a low amino acid diet, *Agric. Biol. Chem.*, 33, 43, 1969.

45. **Said, A. K. and Hegsted, D. M.**, Response of rats to diets of equal chemical essential amino acids, *J. Nutr.*, 100, 1363, 1970.

46. **Cieslak, D. G. and Benevenga, N. J.**, Response of rats to diets of equal chemical score: effect of lysine and threonine as the limiting amino acid and of an amino acid excess, *J. Nutr.*, 166, 969, 1986.

47. **Cieslak, D. G. and Benevenga, N. J.**, The effect of amino acid excess on utilization by the rat of limiting amino acid — threonine, *J. Nutr.*, 114, 1871, 1984.

48. **Cieslak, D. G. and Benevenga, N. J.**, The effect of amino acid excess on utilization by the rat of limiting amino acid — lysine and threonine at equalized food intakes, *J. Nutr.*, 114, 1878, 1984.

49. **Zello, G. A., Pencharz, P. B., and Ball, R. O.**, Phenylalanine flux, oxidation, and conversion to tyrosine in humans studied with L-[1-^{13}C] phenylalanine, *Am. J. Physiol.*, 259, E835, 1990.

50. Protein Requirements, FAO Nutr. Studies No. 16, Food and Agriculture Organization, Rome, 1957.

51. Protein Requirements, FAO/WHO Nutr. Meetings Report, Ser. No. 37, Food and Agriculture Organization, Rome, 1965.

52. **Williams, H. H., Harper, A. E., Hegsted, D. M., Arroyave, G., and Holt, L. E., Jr.**, Nitrogen and amino acid requirements, in *Improvement of Protein Nurture*, U.S. Food and Nutrition Board, National Academy of Science, Washington, D.C., 1974.

53. Energy and Protein Requirements, FAO/WHO, Recommendations by a joint FAO/WHO informal gathering of experts, *Food Nutr.*, 1(2), 11, 1975.

54. **Pellett, P. L. and Young, V. R.**, Role of meat as a source of protein and essential amino acids in human protein nutrition, in *Meat and Health Advances in Meat Research*, Pearson, A. M. and Dutson, T. R., Eds., Elsevier Applied Science, New York, 1990, 329.

55. **Paul, A. A. and Southgate, D. A. T.**, McCance and Widdowson's "The Composition of Foods", 4th ed., Her Majesty's Stationery Office, London, 1978.

56. **McLean, W. C., Klein, G. L., Lopez De Romano, G., Massa, E., and Graham, G. G.**, Protein quality of conventional and high protein rice and digestibility of glutinous and non-glutinous rice by preschool children, *J. Nutr.*, 108, 1740, 1978.

57. *Protein-Energy Requirements Under Conditions Prevailing in Developing Countries: Current Knowledge and Research Needs*, Viteri, R., Whitehead, R., and Young, V. R., Eds., United Nations University, Tokyo, 1979.

58. **Bothwell, T. H., Charlton, R. W., Cook, J. D., and Finch, C. A.**, *Iron Metabolism in Man*, Blackwell Scientific Publishing, Oxford, 1979.

59. **Davies, N. T.**, Anti-nutrient factors affecting mineral utilization, *Proc. Nutr. Soc.* (England), 38, 121, 1970.

60. **Erdman, J. W.**, Oil-seed phytates: nutritional implications, *J. Am. Oil Chem. Soc.*, 56, 736, 1979.

61. **Johnson, N. E., Alcantra, E. N., and Linkswiler, H.**, Effect of level of protein intake on urinary and fecal calcium retention of young adult males, *J. Nutr.*, 100, 1425, 1970.

62. **Margen, S., Chu, J. Y., Kaufman, N. A., and Calloway, D. H.**, Studies in calcium metabolism. I. The calciuretic effect of dietary protein, *Am. J. Clin. Nutr.*, 27, 584, 1974.

63. **Allen, L. H., Oddoye, E. A., and Margen, S.**, Protein-induced hypercalciuria: a long-term study, *Am. J. Clin. Nutr.*, 32, 741, 1979.

64. **Allen, L. H., Barlett, R. W., and Black, G. D.**, Reduction of renal calcium reabsorption in man by consumption of dietary protein, *J. Nutr.*, 109, 1345, 1979.

65. **Spencer, H., Kramer, L., Osis, D., and Norris, C.**, Effect of high protein (meat) intake on calcium metabolism in man, *Am. J. Clin. Nutr.*, 31, 2167, 1978.

66. **Carroll, K. K. and Hamilton, R. M. G.**, Effect of dietary protein and carbohydrate on plasma cholesterol levels in relation to atherosclerosis, *J. Food. Sci.*, 40, 18, 1975.

67. **Hamilton, R. M. G. and Carroll, K. K.**, Plasma cholesterol levels in rabbits fed low fat, low cholesterol diets. Effect of dietary proteins, carbohydrates and fiber from different sources, *Atherosclerosis*, 24, 47, 1976.

68. **Krichevsky, D.**, Vegetable protein and atherosclerosis, *J. Am. Oil Chem. Soc.*, 56, 135, 1979.

69. **Terpstra, A. H. M., West, C. E., Fennis, J. T. C. M., Schouten, J. A., and van de Veen, E. A.,** Hypocholesterolemic effect of dietary soy protein versus casein in rhesus monkeys *(Macaca mulatta), Am. J. Clin. Nutr.,* 39, 1, 1984.

70. *Dietary Proteins, Cholesterol Metabolism and Atherosclerosis,* Suagano, M. and Beynen, A. C., Eds., S. Karger, Basel, 1990.

71. **Terpstra, A. H. M., Hermus, R. J. J., and West, C. E.,** The role of dietary protein in cholesterol metabolism, *World Rev. Nutr. Dietet.,* 42, 1, 1983.

72. **Terpstra, A. H. M., Hermus, R. J. J., and West, C. E.,** Dietary protein and cholesterol metabolism in rabbits and rats, in *Animal and Vegetable Proteins in Lipid Metabolism and Atherosclerosis,* Alan R. Liss, New York, 1983, 19.

73. **Huff, M. W., Giovannetti, P. M., and Wolfe, B. M.,** Turnover of very low-density lipoprotein-apoprotein B is increased by substitution of soybean protein for meat and dairy protein in the diets of hypercholesterolemic man, *Am. J. Clin. Nutr.,* 39, 888, 1984.

74. *Diet, Nutrition and Cancer,* National Research Council, National Academy Press, Washington, D.C., 1982, 478.

75. *Diet and Health: Implications for Reducing Chronic Disease Risk,* NAS/NRC, Committee on Diet and Health, National Research Council, National Academy Press, Washington, D.C., 1989.

Chapter 3

PROTEIN SOURCES FOR USE IN FOOD PRODUCTS

Walter J. Wolf

TABLE OF CONTENTS

I. INTRODUCTION

People of the Western World have a long tradition of consuming a variety of protein foods including meat, milk, fish, eggs, and bread. These foods are well accepted and, indeed, are usually preferred over other protein sources. Except for bread, these protein foods are all of animal origin. Animal-derived foods have the disadvantage of being expensive to produce because of the biological inefficiency of converting plant proteins to animal proteins. It is therefore desirable to consider alternate sources of protein that can be used to extend and imitate the animal protein foods as well as to develop new food products that appeal to consumers on their own merits. The purpose of this section is to examine the total protein supply and then to consider the various sources that can be used to increase the total supply of dietary proteins. Availability, nutritional characteristics, safety considerations, and status for human use are discussed.

II. WORLD PROTEIN SUPPLIES

The human diet is very diverse and man has learned to utilize proteins from a large variety of animal, microbial, and plant sources. Relative amounts of proteins from animals, microorganisms, and plants vary among countries and depend on economics, availability of supplies, cultural preferences, geography, and climate.

A. ANIMAL PROTEINS

These proteins have a long history of use. Consequently they are preferred over microbial and plant proteins in many cultures, especially in developed countries such as the U.S., Canada, and the European nations. In 1984, meat and dairy products each supplied about 23×10^6 metric tons (MT) of protein in world production. Eggs, however, provided 2.7 times as much protein as meat and dairy products combined (Table 1). The fish catch added another 16×10^6 MT of protein for a total of about 185×10^6 MT of animal proteins. However, only about 70% of the total fish is used for edible purpose; thus, the total animal protein available for human consumption is somewhat less.

B. MICROBIAL AND PLANT PROTEINS

World production of microbial proteins in 1984 was only about 0.7×10^6 MT (Table 2); the bulk of this was used for animal feeds. Among the plant proteins, cereals (primarily wheat, rice, maize, and barley) represented about 180×10^6 MT. The other plant materials, including legumes and oilseeds, supplied an additional 70×10^6 MT of proteins. Soybeans are the major protein source in this group and provided 34×10^6 MT. Most soybean protein goes into animal feeds; thus, soybeans are a largely untapped source that could be used for human foods.

III. ALTERNATE PROTEIN SOURCES

In many cultures animal proteins are the preferred forms of dietary protein for esthetic and nutritional reasons. However, their expense makes it difficult to increase dietary protein supplies by producing more animal products. Consideration of alternate protein sources therefore directs attention to microbial and plant proteins. Proteins from alternate sources should have the following characteristics: (1) availability in large supply; (2) low cost; (3) tradition of usage; (4) good nutritional value; (5) safety to humans; (6) good functional properties; and (7) the ability to be incorporated into conventional foods while maintaining traditional quality that consumers expect.

TABLE 1
World Production of Animal Protein Foodstuffs, 1984

	Product (10⁶ Metric tons)	Protein (10⁶ Metric tons)
Meat		
Beef and veal	45.8	8.1
Mutton and lamb	6.1	1.0
Goat	2.0	0.3
Pork	55.5	6.6
Poultry	30.0	6.0
Miscellaneous	4.7	0.8
Subtotal	**144.1**	**22.8**
Dairy products		
Milk, cow, whole	448.6	15.7
Milk, buffalo	31.8	1.3
Milk, sheep	8.3	0.4
Milk, goat	7.5	0.3
Cheese	12.3	2.2
Evaporated condensed milk	4.7	0.4
Dry whole milk	2.0	0.5
Skim milk	4.5	1.6
Dry whey	1.5	0.2
Subtotal	**521.2**	**22.6**
	997.2	123.7
Eggs	83.1	15.6
Fish		
Total	**1,745.6**	**184.7**

Note: Product data from FAO Production Yearbook[1] and FAO Yearbook of Fishery Statistics.[2] Protein production figures were estimated using food composition tables.[3,4]

A. MICROBIAL PROTEINS

Microbial proteins or single-cell proteins (SCP) have been consumed in foods since antiquity in the form of bread, beer, cheese, and yogurt.[7,8] The microbial proteins in these products, however, result from the use of the microorganisms as processing adjuncts and do not supply much protein to the diet (e.g., 1 to 3% yeast protein in bread). Modern technology makes it possible to produce these proteins on a large scale from a variety of substrates.[6,8] These proteins are used primarily for animal feed supplements where cheaper sources of protein, such as soybean meal, are unavailable. Yeasts, with a long history of food consumption and acceptance, are used mainly for their functional properties — flavoring agents and binding of fat and water — instead of as dietary protein. High contents of nucleic acids (6 to 9%) limit the use of yeasts as protein ingredients in foods.[9] Removing nucleic acids by processing would increase production costs which are already high. The future of SCP is uncertain and its past has not been a commercial success story.[6,8,10]

B. PLANT PROTEINS

Among the plant proteins the major potential sources are cereals and oilseeds (Table 2). Other sources, such as green leaves, winged beans, and coconut proteins, have also been investigated and will be considered here.

1. Cereals

Wheat, rice, and corn already are major dietary protein sources in many countries. Significant quantities of corn and sorghum are also used in feeds as energy sources, particularly in the developed countries. Among the cereals, only corn and wheat are processed

TABLE 2
World Production of Microbial and Plant
Protein Foodstuffs, 1984

	Product (10⁶ Metric tons)	Protein (10⁶ Metric tons)
Microbial proteins[6]	1.1	0.7
Cereals		
Wheat	521.7	63.6
Rice	470.0	35.3
Corn (maize)	449.3	42.7
Barley	171.6	18.9
Sorghum	71.7	7.2
Oats	43.4	5.6
Rye	31.1	3.4
Millet	30.9	3.0
Subtotal	**1,789.7**	**179.7**
Potatoes	312.2	6.2
Sweet potatoes	117.3	1.5
Pulses	47.9	10.4
Soybeans	89.9	34.2
Cottonseed	34.2	6.9
Peanuts (in shell)	20.6	4.0
Sunflower seed	15.9	2.0
Rapeseed	16.4	3.9
Sesame	2.1	0.4
Subtotal	**656.5**	**69.5**
Total	**2,447.3**	**249.9**

Note: Product data from FAO Production Yearbook[1] except for microbial proteins. Protein production figures were estimated using food composition tables[3] and other sources.[5]

into protein products. Gluten is isolated from wheat or wheat flour by several methods.[11] Most commonly, flour and water are mixed into a dough, which is then washed to remove the starch. Wheat gluten is used primarily for its unique cohesive and viscoelastic functional properties. A major application is in wheat flours that are low in gluten. In Europe this use eliminates the need to import high-protein wheats from Canada, the U.S., and Australia.[11] Zein is extracted from corn gluten with aqueous alcohol and is used mainly as a coating for nut meats and confections to retard rancidity or moisture penetration. Most cereal proteins are low in lysine; hence they are not suited for use as protein extenders if nutritional quality is important.

2. Soybeans

Historically, soybeans have a long tradition of food use in China and Japan in forms such as tofu (soybean curd), miso, and soy sauce. In the last decade, tofu has also been introduced and popularized in the West. Many supermarkets in the U.S. now sell tofu, and food products containing tofu or tofu-like ingredients are being developed and introduced. Although the tofu market is growing, it is still small compared to the soybean protein ingredients market consisting of flours, concentrates, and isolates. At present, soybeans are the predominant plant source used in the manufacture of edible proteins. Soybean proteins have been commercially available in the U.S. for over 30 years and they are no longer considered new or novel.[12] The soybean protein industry is well established; in the U.S., about a dozen companies manufacture a variety of protein forms. Compositions of these products are given in Table 3. Nutritionally, the essential amino acid compositions of soybean

TABLE 3
Composition of Soybean Protein Products[a]

	Defatted flour (%)[13]	Concentrates (%)		Isolates (%)	
		A[b]	B[c]	A[d]	B[e]
Protein (N × 6.25)	54.7	67.0	66.0	92.8	92.9
Moisture	7.0	6.0	6.0	4.7	7.6
Fat	0.8	0.3	0.3	—	—
Crude fiber	2.4	3.5	3.4	0.1	0.1
Ash	6.0	5.6	4.8	3.8	2.0
Carbohydrates[f]	28.9	17.6	19.5	—	—

[a] As-is basis.
[b] Made by alcohol extraction.[14]
[c] Made by acid leaching.[14]
[d] Sodium proteinate form.[15]
[e] Isoelectric form.[15]
[f] By difference.

proteins compare favorably with recently established Food and Agriculture Organization/World Health Organization/United Nations University (FAO/WHO/UNU) amino acid requirements.[16] Table 4 shows that defatted meal and isolate A meet or exceed all of the amino acid requirements for children and adults. Isolate B meets all requirements except for 2- to 5-year-old children, but it meets 88% of the sulfur amino acid requirement for this group. Two g of soy isolate per 70 kcal of infant formula is more than adequate to meet or exceed the amino acids supplied by human milk at an equivalent caloric intake. Long-term feeding studies with humans support the conclusion that soybean proteins are good sources of essential amino acids. When soybean protein isolates are consumed at 0.8 g/kg body weight per day, the protein nutritional status of young men is maintained adequately.[23] With preschool children, digestibilities, nitrogen balance indices, and plasma protein levels indicated that soybean protein isolates compared adequately with milk.[24] Normal infants fed formulas containing soy protein isolate not fortified with methionine performed less well during the first 6 weeks of life than did breast-fed infants and infants fed milk-based formulas or other soy isolate-based formulas fortified with methionine.[25] The methionine content of soy proteins appears adequate to meet the nutritional need of preschool children and adults, but there may be a need to supplement soy isolate-based formulas for infants. Long-term (6-months) studies of extending beef with soybean protein under practical conditions have revealed no adverse effects on iron and zinc absorption in men, women, and children.[26-28] Consumption by military men or women or by school lunch program participants of beef extended with soybean protein poses no risk of zinc or iron deficiency in these population groups at the levels of consumption used.

Trypsin inhibitors are widely distributed in nature. They occur in many common food items such as legumes, cereal grains, potatoes, and eggs.[29,30] Trypsin inhibitors of soybeans have been studied in great detail because they appear to be involved in the poor nutritional quality exhibited by raw soybeans and affect the pancreas of some animals. Certain animals, particularly rats, show sensitivity of the pancreas to long-term ingestion of trypsin inhibitors. One group of investigators found nodular hyperplasia and alinar adenoma,[31] while another group noted no such changes.[32] Primates, such as cebus monkeys, did not show a pancreatic response when fed soybean protein concentrates or isolates for 2 to 4 years.[33] Normal moist heat processing inactivates most of the soybean trypsin inhibitor activity. Many experts thus agree that properly processed soybean protein products do not pose a hazard to human health at practical levels of consumption.

TABLE 4
FAO/WHO/UNU Amino Acid Requirements and Amino Acid Composition of Plant Proteins (mg/g Protein)

Essential amino acids	Requirements[16] Child, age 2 to 5	Requirements[16] Child, age 10 to 12	Requirements[16] Adult	Soybean Defatted meal[17]	Soybean Isolates A[a]	Soybean Isolates B[b]	Peanut defatted flour[18]	Cottonseed defatted flour[19]	Sunflower kernel protein[c]	Rapeseed defatted meal[21]	Pea flour[d]
Histidine	19	19	16	26	27	22	23	31	28	28	24
Isoleucine	28	28	13	46	49	49	32	32	44	40	40
Leucine	66	44	19	78	81	78	64	60	69	70	72
Lysine	58	44	16	64	64	63	30	49	39	60	74
Methionine + cystine	25	22	27	26	26	22	19	29	26	30	20
Phenylalanine + tyrosine	63	22	19	88	93	93	84	76	85	65	76
Threonine	34	28	9	39	37	37	26	37	37	45	38
Tryptophan	11	9	5	14	15	15	10	21	16	12	11
Valine	35	25	13	46	47	48	53	47	53	51	45

a   EDI-PRO® A, product data sheet, Protein Technologies International, St. Louis, MO.
b   PRO-FAM G-900, product data sheet, Grain Processing Corporation, Muscatine, IA.
c   For "Arrowhead" cultivar.[20]
d   Mean for 16 samples of Century cultivar.[22]

TABLE 5
Functional Properties of Soybean Proteins in Food Systems

Functionality	Mode of action	Food system
Solubility	Protein solvation	Beverages
Water absorption and binding	Hydrogen-bonding of water, entrapment of water (no drip)	Meats, sausages, breads, cakes
Viscosity	Thickening, water binding	Soups, gravies
Gelation	Protein matrix formation and setting	Meats, curds, cheese
Cohesion-adhesion	Protein acts as adhesive	Meats, sausages, baked goods, cheeses, pasta products
Elasticity	Disulfide links in deformable gels	Meats, baked goods
Emulsification	Formation and stabilization of fat emulsions	Sausages, bologna, soups, cakes
Fat adsorption	Binding of free fat	Meats, sausages, doughnuts
Flavor binding	Adsorption, entrapment, release	Simulated meats, baked goods
Foaming	Formation of stable films to entrap gases	Whipped toppings, chiffon desserts, angel food cakes

From Kinsella, J. E., Damodaran, S., and German, B., in *New Food Proteins,* Vol. 5, Seed storage proteins, Altschul, A. M. and Wilcke, H. L., Eds., Academic Press, New York, 1985, 107. With permission.

Ingestion of soy flour may cause flatulence if the level of consumption is high enough. Flatulence is believed to be caused by the oligosaccharides raffinose and stachyose.[34] Defatted soy flours have raffinose plus stachyose contents of 5 to 6%.[35] Processing of defatted flakes into concentrates lowers their oligosaccharide content and flatus activity. Isolates are almost free of oligosaccharides and do not cause flatulence.[34]

A positive nutritional benefit of soybean proteins is their ability to lower serum cholesterol levels in human subjects with hypercholesterolemia. With rabbits, casein is more atherogenic than soy protein isolate and atherogenicity is influenced by the ratio of lysine to arginine of the proteins. Casein exerts its cholesterolemic effect in part by lowering the excretion of cholesterol and lengthening the turnover time. The manner by which single amino acids or peptides formed during protein digestion influence cholesterol metabolism is still unknown.[36] Soybean proteins in their different forms are used in a wide variety of foods, primarily for their functional properties.[13,14,37-39] Infant formulas, dietary wafers, breakfast cereals, and special dietary items are exceptions where soybean protein serves as a dietary protein source. Soybean proteins have an array of functional properties that are important in their use to extend and imitate animal proteins (Table 5). These properties enable food processors to maintain the traditional quality of conventional food products in extended products and to duplicate it in food analogs. For example, the ability to gel and to bind fat and water is necessary in processed meat applications (cured meats). Solubility is crucial in products such as milk and milk analogs.

Defatted soybean flour finds its major application in baked items where the blending with cereals increases both the quantity and quality of the protein in the cereal products. Soy protein concentrates and isolates are used extensively in processed meats, meat analogs, poultry, and fish products. Surimi-based seafood items are potential outlets for soybean proteins. A recent review discusses the contributions of soybean proteins to the texture of gelled muscle protein foods derived from meat and fish.[40] For dairy products and analogs, isolates are the most suitable protein form to use. Soybean protein isolate has been used in infant formulas for over 25 years.[41] This application is very critical because soybean is the sole source of protein in these products. Soy-based formulas are generally fed when children develop allergies to cow milk. Direct food use of soybean proteins in the U.S. amounts to only about 2 to 3% of the crop. Hence, increased conversion of defatted flakes to edible protein products would not likely have an adverse effect on the supply of defatted meal for feeds.[39]

3. Peanuts

Processes for the manufacture of partially defatted peanuts, defatted flours, concentrates, and isolates have been developed.[42-45] However, only partially defatted peanut and flours are produced in the U.S. at present. Partially defatted peanuts are made by hydraulic pressing to remove 50 to 60% of the oil, reconstituting by soaking in salt solution, and then deep-fat frying. One company makes partially defatted peanut flours by hydraulic pressing without heat. These products contain 40 to 42% fat and 32 to 34% protein and are available with varying degrees of roasting. Partially defatted flours are used extensively in granola bars for peanut butter-flavored coatings and in other confections to provide peanut flavor. Edible-grade defatted peanut flours and grits were available commercially until about 1986 and contained about 57% crude protein.[45] A peanut protein isolate was commercially available in India until 1985. Uses included toned (extended) cow and buffalo milks and baby foods.[42] In the U.S. the high price of peanuts has limited the production and use of peanut proteins.

Peanut proteins are low in lysine, leucine, sulfur amino acids, and threonine. Consequently, they do not meet amino acid requirements for certain age groups (Table 4). As with cottonseed, peanuts are subject to infection by the mold *Aspergillus flavus* if they are improperly handled and stored. The permitted level of aflatoxins in peanut products in the U.S. is 20 ppb; the average level in peanut products is 2 ppb.[46] Peanuts also contain oligosaccharides of the raffinose family that cause flatulence. An analysis of defatted peanut meal indicated 0.33% raffinose, 0.99% stachyose, and a trace of verbascose; by comparison, raffinose plus stachyose in soy flours is 5 to 6%.[35]

4. Cottonseed

Proflo, a defatted cottonseed flour prepared by screw pressing, became available in the U.S. in 1939, but production of the edible-grade product ended in 1975. An industrial grade of the product is still manufactured for use in fermentation media. Proflo, made from glanded cottonseed, was used in baked goods to provide functional properties such as dough machinability, controlled spread, reduced fat absorption, and improved browning.[47-49] A Proflow-like flour is made by a company in Israel for domestic use and export. It is used in baked goods, chocolate substitutes, snacks, breakfast cereals, and textured vegetable protein products. Specifications for the product include a maximum of 0.05% (500 ppm) of free gossypol. This compares with a limit of 450 ppm imposed by the U.S. Food and Drug Administration and 600 ppm set by the Protein Advisory Group of the FAO and WHO.

Gossypol is a yellow polyphenolic compound that is toxic to some monogastric animals; toxicity in humans has not been reported. Glandless cotton varieties that lack gossypol have been developed. Defatted flours made from glandless seed were introduced for market development in the U.S. in 1986, but the company ceased operations in early 1988. At present there are no edible cottonseed protein products produced in the U.S. Other anti-nutritional factors found in cottonseed flour include the cyclopropenoid fatty acids, malvalic and sterculic acids. In rainbow trout these compounds act as cocarcinogens in the presence of aflatoxins;[50] their effects in humans are unknown. Cottonseed is prone to infection by *A. flavus* and contamination by aflatoxin; hence seed intended for human consumption must be stored carefully. Compared to the FAO/WHO/UNU amino acid requirements, cottonseed flour is low in lysine and leucine for 2- to 5-year-old children, but adequate for older children and adults (Table 4).

5. Sunflower Seed

Pilot plant preparations of sunflower seed flour, protein concentrates, and isolates have been available in the U.S., Italy, and Canada in recent years. Nevertheless, a viable food protein ingredients industry for this protein source has yet to be established. Difficulties in dehulling (hulls are high in fiber) and discoloration caused by oxidation of phenolic acids continue to limit development. Because of the phenolic acids, isolates prepared from defatted flours by alkali extraction followed by acid precipitation are green in color. Sunflower seeds of the "confectionery" type are available as "in-shell" and dehulled forms. The kernels are either roasted and consumed as snacks or left unroasted. They are used as cooking ingredients, in mixtures of dried fruits and nuts, and as garnishes for salads. Other uses of the kernels are in breads, buns, and breakfast cereals.[51]

Sunflower proteins meet the FAO/WHO/UNU amino acid requirements except for lysine for children; for adults the lysine level exceeds requirements (Table 4). Sunflower seeds also contain oligosaccharides. Defatted meal contains 3.1% raffinose and 0.14% stachyose as determined by high-performance liquid chromatography.[35]

6. Rapeseed

Extensive research and development in Canada and Sweden have led to significant progress in improving the composition of rapeseed. In the 1970s Canadian breeders introduced varieties that were low in glucosinolate and in erucic acid contents. These new varieties are collectively known as canola to distinguish them from the older high erucic acid-type rapeseed. Canola varieties have replaced all of the rapeseed grown in Canada and a similar changeover was anticipated for Europe by 1990.[5] Although glucosinolate levels of canola are much lower than in conventional rapeseed, present levels (30 μmol or less of glucosinolates per gram of dry, oil-free meal) are still too high for direct food use. As a result, present canola varieties are not used in the form of edible flours and grits.

Rapeseed meals also contain phenolic compounds, such as sinapine, which taste bitter and cause discoloration by reacting with other components. Oxidized phenolic compounds react with proteins, thus affecting their nutritional availability. Meals also contain up to 35% by weight of hulls. After grinding, they occur as dark particles in an otherwise light meal. Such particles make meal undesirable in appearance for food uses. Glucosinolates, phenolic compounds, and hulls must be removed to make canola meal acceptable for food uses. Notwithstanding these negative factors, rapeseed has an excellent amino acid composition. The meal meets all of the FAO/WHO/UNU requirements for children and adults (Table 4). Rapeseed meals can be upgraded by conversion to protein concentrates. Similar processes for protein concentrates have been developed by the Karlshamms company in Sweden[5] and the Food Research Institute in Canada.[52] In both processes the seed is dehulled, heated to inactivate enzymes that are capable of hydrolyzing the glucosinolates, and leached with water to remove the glucosinolates and other water-soluble compounds. Next the seed is dried, extracted with hexane to remove the oil, and then ground to the desired particle size. The resulting concentrates contain up to 65% protein (N × 6.25) and are essentially free of glucosinolates and phenolic compounds. Nutritional testing indicates that the concentrates have a protein efficiency ratio equal to or better than that of casein. The apparent toxicity of rapeseed protein concentrate in pregnant rats has been attributed to a zinc deficiency that is overcome by supplementing with zinc.[52] Zinc deficiency of the concentrates is attributed to high phytate contents (5 to 7.5%). Attempts to prepare protein isolates from rapeseed have been less successful. Yields were low because much of the protein did not precipitate when an alkaline extract was acidified as in the usual procedure for preparing soybean protein isolates.[53] The phenolic compounds also cause discoloration of the proteins. At present, economics do not favor the manufacture of either rapeseed protein concentrates or isolates. Consequently, they are not available commercially.

7. Leaves

Extensive research and development has been conducted in different countries to recover proteins from green leaves which are the largest world supply of protein. The high fiber content of leaves requires that the proteins be separated and concentrated. A variety of processes for concentrating the protein have been investigated.[54-56] Nonetheless, there is no commercial production of edible-grade leaf protein at present. Near-term prospects for establishment of such an industry are not encouraging. Major problems that need to be solved include consumer acceptance (texture, green color, and grassy flavor), toxicity (presence of compounds such as saponins and pheophorbide *a*, which is a photosensitizing agent), and economics (must be competitive with soybean and other oilseed meals).

8. Peas

In the 1970s, Canadian scientists at the Prairie Regional Laboratory in Saskatoon, Saskatchewan developed processes for the production of pea flours and pea protein concentrates by air classification.[57] These products were available commercially beginning in 1976, but production ceased in 1985. Another Canadian company manufactures pea protein concentrates (83 to 85% protein) as a by-product in the production of pea starch by a wet process. The concentrates are used in the baking industry as replacers for nonfat dry milk and as protein supplements; another outlet is veal calf feeding. Pea protein has a reasonably good amino acid composition except for the sulfur amino acids. Methionine plus cystine contents are about 80% of the FAO/WHO/UNU requirement for 2- to 5-year-old children, but meet requirements for adults (Table 4). Economic factors will determine the future viability of the edible pea protein industry. Soybeans are the major competitor for pea proteins in Canada.

9. Coconuts

Experimental coconut protein products are reported in the literature. Coconut flour is made by grinding coconut meat after it is dried and the oil is extracted. These products will contain 20 to 24% protein and 8 to 10% crude fiber. Spray drying of coconut milk after first removing the oil by centrifugation yields a white powder containing 32% protein and only 0.2% crude fiber. This product is called Cocopro. Coconut flour has a good nutritional value and does not contain any known antinutritional factors. The future for coconut proteins is uncertain. There is no commercial production, product specifications have not been developed, and regulatory agencies have not examined the products. In 1979 it was estimated that potential edible coconut protein supplies were only about 2% of potential edible soybean supplies.[58]

10.. Other Legumes

Several legumes besides soybeans and peanuts supply substantial amounts of dietary protein, especially to residents of rural Africa and Asia. These legumes include chick-peas,[59] pigeon peas,[60] common dry beans, *Phaseolus vulgaris*,[61,62] black gram,[63] lupines,[64] and faba beans.[65] Most of these legumes are consumed directly as foods and are not processed into food ingredients. Many of them are high in carbohydrate content (mainly starch) and low in fat. Consequently, even defatted flours are low in protein compared to the defatted oilseed meals. Protein concentrates and isolates have been made experimentally from many of them, but there is no large-scale commercial processing.

Faba bean protein isolate was manufactured and used commercially for a time in the U.K. to produce a spun fiber, but its production was discontinued because of unfavorable economics. Commercial production of air-classified faba bean protein concentrate and starch is reported in France. Sweet lupines (varieties low in alkaloid contents) are under development in the U.S. A company in Minnesota has available hulled and dehulled flours plus a dietary fiber flour made by grinding the hulls. Winged bean, *Psophocarpus tetragonolobus,* has also received the attention of research workers.[66,67] Protein and oil contents of some varieties approach those of soybeans. Major obstacles to development of a protein industry based on this legume are a lack of large-scale production and an industry to process winged beans into oil and meal. At present, the winged bean is mainly a backyard garden crop in the tropics and the pods are often consumed as a green vegetable. Present varieties are climbing herbaceous perennials grown on poles, trellises, or fences. They are not suited for mechanized cultivation on a large scale.

IV. CONCLUSIONS

A survey of alternative protein sources reveals a relatively small number of possibilities. At present, SCP have limited potential except for specialty uses such as flavorings, where they command a high price. Their nucleic acid content limits the use of such products at high levels. The limit for nucleic acid intake is 2 g/d.

Of the plant proteins, soybeans are the clear choice of an alternative source of proteins for use as food ingredients. The industry is well established and a wide range of products — isolates, concentrates, and flours in untextured and textured forms — is available. The products range in protein content from 90% for the isolates down to 50% for the defatted flours. The industry also has considerable technical information on applications of soy proteins in a wide array of foods including baked goods, confections, processed meats, meat analogs, coffee whiteners, infant formulas, milk replacers, and beverages. Because of their good balance of essential amino acids, soy proteins can extend or simulate animal protein products while maintaining traditional nutritional quality. Of the soy protein forms available, isolates are the most highly refined and therefore the most expensive. Isolates, however, are the most versatile because both the soluble and insoluble carbohydrates of flours are removed in conversion to the isolates.

Wheat gluten and zein from corn occupy specialty markets because of their unusual functional properties. However, their low lysine contents make them unsuitable, from a nutritional standpoint, for extending or replacing animal proteins.

Other plant proteins used commercially — peanuts, cottonseed, sunflower seed, peas, and lupines — are produced on a much smaller scale. Many are available only in one or two forms, e.g., sunflower proteins are available only in the form of the kernels, and cottonseed proteins only as defatted flours. Peanuts are too expensive to compete with soybeans and do not have a good balance of essential amino acids. Cottonseed still faces uncertainties of safety (gossypol and cyclopropenoid fatty acids) and sunflower proteins have problems of quality (high fiber in flours and discoloration caused by phenolic acids).

Of the remaining plant protein sources, rapeseed, leaf proteins, coconut, winged bean, and other legumes are potential sources. However, near-term prospects are not very encouraging. Uncertainties about their future development include problems of safety (glucosinolates in rapeseed and saponins plus pheophorbide a in leaf proteins), quality (high fiber in coconut and green color of leaf proteins), availability (winged bean and coconut), and economics (all must be able to compete with soybeans).

REFERENCES

1. FAO Production Yearbook, Vol. 38, Food and Agriculture Organization, Rome, 1985.
2. Yearbook of Fishery Statistics, Fishery Commodities, Vol. 61, Food and Agriculture Organization, Rome, 1987.
3. Amino-Acid Content of Foods and Biological Data on Proteins, FAO, Nutritional Studies No. 24, Food and Agriculture Organization, Rome, 1970.

4. **Posati, L. P. and Orr, M. L.,** Composition of Foods, Dairy and Egg Products—Raw, Processed, Prepared, Agriculture Handbook No. 8-1, Agricultural Research Service, U.S. Department of Agriculture, Washington, D.C., 1976.

5. **Ohlson, R.,** Rapeseed, in *New Protein Foods,* Vol. 5, Seed storage proteins, Altschul, A. M. and Wilcke, H. L., Eds., Academic Press, New York, 1985, 339.

6. **Tuse, D.,** *Single-Cell Protein: Current Status and Future Prospects, Crit. Rev. Food Sci. Nutr.,* 1984, 19, 273.

7. **Litchfield, J. H.,** Microbial cells on your menu, *Chemtech, Technol.,* 8, 218, 1978.

8. **Litchfield, J. H.,** Single-cell proteins, *Science,* 219, 740, 1983.

9. **Litchfield, J. H.,** Foods, nonconventional, *Encycl. Chem. Technol.,* 11, 184, 1980.

10. **Sherwood, M.,** The case of the money-hungry microbe, *Bio/Technology,* 2, 606, 1984.

11. **Schofield, J. D. and Booth, M. R.,** Wheat proteins and their technological significance, in *Developments in Food Proteins,* Vol. 2, Hudson, B. J. F., Ed., Applied Science Publishers, Barking, England, 1983, 1.

12. **Wolf, W. J.,** Progress and future needs for research in soya protein utilization and nutrition, *J. Am. Oil Chem. Soc.,* 58, 467, 1981.

13. **Smith, A. K. and Circle, S. J.,** *Soybeans: Chemistry and Technology,* Vol. 1, *Proteins,* AVI Publishing, Westport, CT, 1972.

14. **Campbell, M. F., Kraut, C. W., Yackel, W. C., and Yang, H. S.,** Soy protein concentrate, in *New Protein Foods,* Vol. 5, Seed storage proteins, Altschul, A. M. and Wilcke, H. L., Eds., Academic Press, New York, 1985, 301.

15. **Meyer, E. W.,** Oilseed protein concentrates and isolates, *J. Am. Oil Chem. Soc.,* 48, 484, 1971.

16. Energy and Protein Requirements, Report of a Joint FAO/WHO/UNU Expert Consultation, Tech. Rep. Ser. No. 724, World Health Organization, Geneva, 1985.

17. **Cavins, J. F., Kwolek, W. F., Inglett, G. E., and Cowan, J. C.,** Amino acid analysis of soybean meal: interlaboratory study, *J. Assoc. Off. Anal. Chem.,* 55, 686, 1972.

18. **Cater, C. M. and Rhee, K. C.,** Protein concentrates and isolates, in *Peanut Production in Texas,* Research Monograph, RM-3, Agricultural Experiment Station, College Station, TX, 1975, 115.

19. **Zarins, Z. M. and Cherry, J. P.,** Storage proteins of glandless cottonseed flour, *J. Food Sci.,* 46, 1855, 1981.

20. **Robinson, R. G.,** Amino acid and elemental composition of sunflower and pumpkin seeds, *Agron. J.,* 67, 541, 1975.

21. **Clandinin, D. R.,** *Canola Meal for Livestock and Poultry,* Publ. No. 59, Canola Council of Canada, Winnipeg, 1981.

22. **Holt, N. W. and Sosulski, F. W.,** Amino acid composition and protein quality of field peas, *Can. J. Plant Sci.,* 59, 653, 1979.

23. **Young, V. R., Wayler, A., Garza, C., Steinke, F. H., Murray, E., Rand, W. M., and Scrimshaw, N. S.,** A long-term metabolic balance study in young men to assess the nutritional quality of an isolated soy protein and beef proteins, *Am. J. Clin. Nutr.,* 39, 8, 1984.

24. **Torun, B.,** Nutritional quality of soybean protein isolates: studies in children of preschool age, in *Soy Protein and Human Nutrition,* Wilcke, H. L., Hopkins, D. T., and Waggle, D. H., Eds., Academic Press, New York, 1979, 101.

25. **Fomon, S. J., Ziegler, E. E., Filer, L. J., Jr., Nelson, S. E., and Edwards, B. B.,** Methionine fortification of a soy protein formula fed to infants, *Am. J. Clin. Nutr.,* 32, 2460, 1979.

26. **Miles, C. W., Bodwell, C. E., Morris, E., Ziyad, J. A., Prather, E. S., Mertz, W., and Canary, J. J.,** Long-term consumption of beef extended with soy protein by men, women and children. I. Study design, nutrient intakes, and serum zinc levels, *Plant Foods Hum. Nutr.,* 37, 341, 1987.

27. **Bodwell, C. E., Miles, C. W., Morris, E., Prather, E. S., Mertz, W., and Canary, J. J.,** Long-term consumption of beef extended with soy protein by men, women and children. II. Effects on iron status, *Plant Foods Hum. Nutr.,* 37, 361, 1987.

28. **Morris, E. R., Bodwell, C. E., Miles, C. W., Mertz, W., Prather, E. S., and Canary, J. J.,** Long-term consumption of beef extended with soy protein by children, women and men: III. Iron absorption by adult men, *Plant Foods Hum. Nutr.,* 37, 377, 1987.

29. **Doell, B. H., Ebden, C. J., and Smith, C. A.,** Trypsin inhibitor activity of conventional foods which are part of the British diet and some soya products, *Plant Foods Hum. Nutr.,* 31, 139, 1981.

30. **Rackis, J. J., Wolf, W. J., and Baker, E. C.,** Protease inhibitors in plant foods: content and inactivation, *Adv. Exp. Biol. Med.,* 199, 299, 1986.

31. **Gumbmann, M. R., Spangler, W. L., Dugan, G. M., Rackis, J. J., and Liener, I. E.,** The USDA trypsin inhibitor study. IV. The chronic effects of soy flour and soy protein isolate on the pancreas in rats after two years, *Plant Foods Hum. Nutr.,* 35, 275, 1985.

32. **Richter, B. D. and Schneeman, B. O.,** Pancreatic response to long-term feeding of soy protein isolate, casein or egg white in rats, *J. Nutr.,* 117, 247, 1987.

33. **Ausman, L. M., Harwood, J. P., King, N. W., Sehgal, P. K., Nicolosi, R. J., Hegsted, D. M., Liener, I. E., Donatucci, D., and Tarcza, J.,** The effects of long-term soy protein and milk protein feeding on the pancreas of *Cebus albifrons* monkeys, *J. Nutr.,* 115, 1691, 1985.

34. **Rackis, J. J.,** Flatulence caused by soya and its control through processing, *J. Am. Oil Chem. Soc.,* 58, 503, 1981.

35. **Kuo, T. M., Van Middlesworth, J. F., and Wolf, W. J.,** Content of raffinose oligosaccharides and sucrose in various plant seeds, *J. Agric. Food Chem.,* 36, 32, 1988.

36. **Kritchevsky, D., Tepper, S. A., and Klurfeld, D. M.,** Dietary protein and atherosclerosis, *J. Am. Oil Soc.,* 64, 1167, 1987.

37. **Kinsella, J. E., Damodaran, S., and German, B.,** Physicochemical and functional properties of oilseed proteins with emphasis on soy proteins, in *New Food Proteins,* Vol. 5, Seed storage proteins, Altschul, A. M. and Wilcke, H. L., Eds., Academic Press, New York, 1985, 107.

38. **Kolar, C. W., Richert, S. H., Decker, C. D., Steinke, F. H., and Vander Zanden, R. J.,** Isolated soy protein, in *New Protein Foods,* Vol. 5, Seed storage proteins, Altschul, A. M. and Wilcke, H. L., Eds., Academic Press, New York, 1985, 259.

39. **Sipos, E. F.,** Edible uses of soybean protein, in *Soybean Utilization Alternatives,* McCann, L., Ed., Center for Alternative Crops and Products, University of Minnesota, St. Paul, 1988, 57.

40. **Foegeding, E. A. and Lanier, T. C.,** The contribution of nonmuscle proteins to texture of gelled muscle protein foods, *Cereal Foods World,* 32, 202, 1987.

41. **Sarett, H. P.,** Soy-based infant formulas, in *World Soybean Research, Proceedings of the World Soybean Research Conference,* Champaign, IL, Hill, L. D., Ed., Interstate Printers & Publishers, Danville, IL, 1976, 840.

42. **Natarajan, K. R.,** Peanut protein ingredients: preparation, properties, and food uses, *Adv. Food Res.,* 26, 215, 1980.

43. **Rhee, K. C.,** Peanuts (groundnuts), in *New Protein Foods,* Vol. 5, Seed storage proteins, Altschul, A. M. and Wilcke, H. L., Eds., Academic Press, New York, 1985, 359.

44. **Conkerton, E. J. and Ory, R. L.,** Peanuts as food proteins, in *Developments in Food Proteins,* Vol. 5, Hudson, B. J. F., Ed., Elsevier Applied Science, London, 1987, 1.

45. **Ayres, J. L., Branscomb, L. L., and Rogers, G. M.,** Processing of edible peanut flour and grits, *J. Am. Oil Chem. Soc.,* 51, 133, 1974.

46. **Lusas, E. W.,** Food uses of peanut protein, *J. Am. Oil Chem. Soc.,* 56, 425, 1979.

47. **Lusas, E. W. and Jividen, G. M.,** Glandless cottonseed: a review of the first 25 years of processing and utilization research, *J. Am. Oil Chem. Soc.,* 64, 839, 1987.

48. **Lusas, E. W. and Jividen, G. M.,** Characteristics and uses of glandless cottonseed food protein ingredients, *J. Am. Oil Chem. Soc.,* 64, 973, 1987.

49. **Frank, A. W.,** Food uses of cottonseed protein, in *Developments in Food Proteins,* Vol. 5, Hudson, B. J. F., Ed., Elsevier Applied Science, London, 1987, 31.

50. **Hendricks, J. D., Sinnhuber, R. O., Loveland, P. M., Pawlowski, N. E., and Nixon, J. E.,** Hepatocarcinogenicity of glandless cottonseeds and cottonseed oil to rainbow trout *(Salmo gairdnerii), Science,* 208, 309, 1980.

51. **Lusas, E. W.,** Sunflower seed protein, in *New Protein Foods,* Vol. 5, Seed storage proteins, Altschul, A. M. and Wilcke, H. L., Eds., Academic Press, New York, 1985, 393.

52. **Jones, J. D.,** Rapeseed protein concentrate preparation and evaluation, *J. Am. Oil Chem. Soc.,* 56, 716, 1979.

53. **Sosulski, F. W.,** Rapeseed protein for food use, in *Developments in Food Proteins,* Vol. 2, Hudson, B. J. F., Ed., Applied Science Publishers, London, 1983, 109.

54. **Kohler, G. O. and Knuckles, B. E.,** Edible protein from leaves, *Food Technol.,* 31(5), 191, 1977.

55. **Fiorentini, R. and Galoppini, C.,** The proteins from leaves, *Plant Foods Hum. Nutr.,* 32, 335, 1983.

56. **Telek, L. and Graham, H. D.,** Leaf protein concentrates, AVI Publishing, Westport, CT, 1983.

57. **Sosulski, F. W.,** Legume protein concentration by air classification, in *Developments in Food Proteins,* Vol. 2, Hudson, B. J. F., Ed., Applied Science Publishers, London, 1983, 173.

58. **Hagenmaier, R.,** Experimental coconut protein products, *J. Am. Oil Chem. Soc.,* 56, 448, 1979.

59. **Chavan, J. K., Kadam, S. S., and Salunkhe, D. K.,** Biochemistry and technology of chickpea (Cicer arietinum L.) seeds, *Crit. Rev. Food Sci. Nutr.,* 25, 107, 1986.

60. **Salunkhe, D. K., Chavan, J. K., and Kadam, S. S.,** Pigeonpea as an important food source, *Crit. Rev. Food Sci. Nutr.,* 23, 103, 1986.

61. **Sathe, S. K., Deshpande, S. S., and Salunkhe, D. K.,** Dry beans of *Phaseolus,* a review. Part 1. chemical composition: proteins, *Crit. Rev. Food Sci. Nutr.,* 20, 1, 1984.

62. **Sgarbieri, V. C. and Whitaker, J. R.,** Physical, chemical, and nutritional properties of common bean (*Phaseolus*) proteins, *Adv. Food Res.,* 28, 93, 1982.

63. **Reddy, N. R., Salunkhe, D. K., and Sathe, S. K.,** Biochemistry of black gram (*Phaseolus mungo L.*), a review, *Crit. Rev. Food Sci. Nutr.,* 16, 49, 1982.

64. **Aguilera, J. M. and Trier, A.,** The revival of the lupin, *Food Technol.,* 32(8), 70, 1978.
65. **Bramsnaes, F. and Sejr Olsen, H.,** Development of field pea and faba bean proteins, *J. Am. Oil Chem. Soc.,* 56, 450, 1979.
66. **Sri Kantha, S. and Erdman, J. W., Jr.,** The winged bean as an oil and protein source, a review, *J. Am. Oil Chem. Soc.,* 61, 515, 1984.
67. **Kadam, S. S. and Salunkhe, D. K.,** Winged bean in human nutrition, *Crit. Rev. Food Sci. Nutr.,* 21, 1, 1984.

Chapter 4

CHEMICAL COMPOSITION AND CONTENT OF POTENTIALLY HAZARDOUS CONTAMINANTS IN ISOLATED SOY PROTEIN AND SOY CONCENTRATES

Vadim G. Vysotsky, Victor A. Tutelyan, and Valentin M. Zhminchenko

Despite the consistent increase in agricultural, fish, and traditional food production there continue to be inadequate amounts of protein and high-quality proteins in some parts of the world. Increasing the production of this essential and economically important dietary component is one of the important medical and social factors for all societies. The solution to the limitation of adequate protein and essential amino acid supplies needs to be resolved by a multifaceted approach including the expansion of tradition food sources by employing improved technological methods and by using new food protein sources to expand the available protein.[1-4] This latter approach holds promise for exploiting the potential of protein resources in the biosphere for nutritional purposes. Conventional methods of producing food often result in large quantities of waste and loss of potential protein.[1,5,6] Bypassing the traditional food chains and directly using nontraditional proteins in the human diet requires finding resolutions for a number of problems. These problems include the issues related to safety, protein value (nutritional), functional suitability as food ingredients, organoleptic acceptability for the consumer, and economical costs of production and delivery.

Vegetable protein foods are a large potential protein source which can be used for the human population. The traditional use of these protein sources for animals and then consuming the animal protein products results in large losses in potential protein. In addition, the capital investment in animal production facilities can be prohibitive in many societies. Therefore, direct use of plant proteins has the potential for conservation of protein and minimizes economic investment to increase production.

Soy protein products, from a practical standpoint, offer a very large available protein source for the human diet. Soybean products have a history of production and consumption by both animals and humans. The availability of new improved isolated soy protein and concentrates makes these ingredients acceptable for use in human foods from an organoleptic and functional standpoint. Therefore, soy protein products were selected for evaluation as the most appropriate new protein source available to the Soviet food system.

The first step in evaluating new protein sources for the human diet is an examination of the chemical composition of protein source.[7,8] This evaluation should include evaluation of the amino acid content and the content of foreign and potentially detrimental factors.

Four isolated soy proteins (PP500E, PP710, Protein Technologies International, U.S., Ardex-M, Ardex-SP6, ADM®, U.S.) and two types of soy concentrates (Unico®, Unilever Co., Netherlands; VNIIZH, All-Union Scientific Research Institute of Fats, U.S.S.R.) were evaluated. The proximate analysis of these six soy products was determined using standard analytical methods. The moisture was determined by drying samples at a temperature of 133°C for a period of 2 h.[9] The protein content was evaluated according to the Kjeldahl method using a Kjeltek® system III "1030-auto" analyzer manufactured by the Tekator Company (Sweden) and drawing upon a coefficient for recalculating the overall nitrogen in protein equal to 6.25.[10] The level of fat was determined by extraction with chloroform-methanol.[11] The content of ash was analyzed after burning the samples in a muffle furnace at a temperature of 600°C.[12] Carbohydrate content was determined by difference.

The amino acid composition of the isolated soya protein and concentrates was determined with the help of three analytical methods for the purpose of obtaining a complete aminogram

TABLE 1
Chemical Composition of Isolated Soy Proteins and
Soy Concentrates

Product	Moisture	Percent on a dry-matter basis			
		Protein	Fat	Ash	Carbohydrate
PP500E	5.5	91.5	1.0	4.0	3.5
PP710	5.5	91.5	0.9	4.0	3.6
Ardex-M	4.6	88.6	1.8	3.7	5.9
Ardex-SP6	5.4	87.5	1.8	4.9	5.8
Unico®	8.0	68.9	0.4	6.2	24.5
VNIIZH	8.3	69.0	0.5	5.2	25.3

for the proteins. The content of most amino acids was examined via ion exchange chromatography using the method developed by Moore and Stein.[13] The concentration of cystine was evaluated after acidifying it with performic acid until cysteic acid was obtained.[14] The content of tryptophan was determined by the method of Spies[16] and Spies and Chambers[15] using alkaline hydrolysis.

The content of heavy metals in the products was determined as well. Colorimetry was employed in analyzing the arsenic,[17] while cadmium, lead, and mercury were determined by the method of atomic absorption spectrophotometry.

The content of trypsin inhibitor was evaluated using the method of Kakade et al.[18] The aflatoxin content was measured by means of the method developed for international application.[19]

In Table 1 are presented the proximate analysis for the six soy products evaluated in this study. As can be seen from the data on the table, differences in protein content were observed within the group of isolated soya proteins which varied by 4.0%. It must also be underscored that the content of fat in the isolated soya proteins from the ADM® company was twice as high as that found in the same types of products made by Protein Technologies International.

The level of protein in both types of concentrates was virtually the same. At the same time, the soya protein concentrates had the lowest content of fats of all products examined, which could be an advantage if these products are stored for long periods. However, the fat content of all of the soya protein products is low, which allows them to be stored for long periods.

The essential amino acid composition in the isolated soya proteins and concentrates is summarized in Table 2. It is clear from the data in this table that all soya protein products were virtually identical, both with respect to the levels of individual essential amino acids as well as to the total essential amino acid content. The possible exception to this rule were the "Ardex" isolated soya proteins, whose tryptophan content was lower. This could become restricting for protein compositions which include this latter protein product.

The results obtained from examinations carried out to determine the content of several foreign substances in the isolated soya proteins and concentrates are presented in Table 3. An analysis of the data reveals that there is a certain scattering of values for the content of several contaminants and antinutritional factors in the soya products. The highest values for the content of the trypsin inhibitor were found in the Ardex-M isolated soya protein and Unico® soya protein concentrate. It must be emphasized that this set of facts cannot serve as a barrier to using certain quantities of these products in combined food products. However, using the latter protein preparations as the sole source of nitrogen in the human diet — such

TABLE 2
Essential Amino Acid Content of Isolated Soy Protein and Soy Concentrates

Amino acids g/100 g protein	Soy protein products					
	Isolates				Concentrates	
	PP500E	PP710	Ardex-M	Ardex-SP6	Unico®	VNIIZH
Isoleucine	4.9	4.9	4.9	4.8	5.2	5.1
Leucine	7.8	8.0	8.3	8.1	7.6	7.8
Lysine	6.3	6.3	6.2	6.2	6.2	7.1
Methionine + cystine	2.6	2.6	2.7	2.7	2.8	2.6
Phenylalanine + tyrosine	9.0	9.0	8.5	8.5	8.7	8.5
Threonine	3.8	3.8	4.0	3.9	4.0	3.8
Tryptophan	1.4	1.4	1.0	1.0	1.5	1.6
Valine	5.0	5.0	5.3	5.0	5.1	5.2
Total essential amino acids	40.8	41.0	40.9	40.2	41.1	41.7

TABLE 3
Content of Foreign Substances in Isolated Soy Protein and Soy Concentrates

Substance	Soy protein products					
	Isolates				Concentrates	
	PP500E	PP710	Ardex-M	Ardex-SP6	Unico®	VNIIZH
Arsenic (mg/kg)	<0.2	<0.2	<1.0	<1.0	<1.0	<1.0
Lead (mg/kg)	<0.5	<0.5	<0.5	<0.5	<2.0	<1.0
Mercury (mg/kg)	<0.03	<0.03	<0.1	<0.1	<0.03	<0.1
Cadmium (mg/kg)	<0.2	<0.2	<0.2	<0.2	<0.2	<0.2
Trypsin inhibitor (TI/kg)	10.0	5.0	30.0	20.0	20.0	5
Aflatoxin (μg/kg)	<2.0	<2.0	<5.0	<5.0	<3.0	<2.0

as in products for enteric diets and infant formulas — may pose a number of problems requiring biomedical examinations. Data obtained from a chemical analysis show the isolated soya protein PP710 to be the best suited for use in diets as the sole protein source.

In conclusion, it is important to stress that the results from studies of the chemical composition of a list of soya protein isolates and concentrates outlined herein may only be viewed as initial data to be used in examinations of their safety, biological value, and digestibility, and do not constitute a basis for deciding whether or not these products may be used for nutritional purposes.

REFERENCES

1. **Tolstoguzov, V. B.,** *Synthetic Foodstuffs,* Nauka, Moscow, 1978.
2. **Ferrando, R.,** *Traditional and Non-Traditional Foods,* Food and Nutrition Series N2, Food and Agriculture Organization, Rome, 1981.
3. **Pokrovski, A. A.,** *Single-cell Proteins (A Review),* Main Administration for Microbiological Industry, SONTI, Moscow, 3, 1971.
4. **Hardin, C. M.,** Impact of plant proteins in worldwide food systems, in *Soy Protein and Human Nutrition,* Wilcke, H. L., Hopkins, D. T., and Waggle, D. H., Eds., Academic Press, New York, 1979, 5.
5. **Rgadchikov, V. G.,** *The Improvement of Grain Proteins and Their Evaluation,* Kolos, Moscow, 1978, 19.

6. **Pimentel, D., Oltenaeu, P. A., Nesheim, M. C., Krummel, J., and Allen, M. S.,** The potential for grass-fed livestock: resource constraints, *Science,* 207, 843, 1980.
7. Pre-clinical testing of novel sources of protein, *PAG Guidelines N6,* PAG Bulletin, IV, 1974, 3.
8. Pre-clinical testing of novel sources of food, Food and Nutrition Bulletin, *PAG/UNU Guidelines N6,* UNU, I, 5, 1983, 60.
9. AOAC, *Official Methods of Analysis,* 14th ed., Moisture in feeds, 7.007, Association of Official Analytical Chemists, Arlington, VA, 1984.
10. AOAC, *Official Methods of Analysis,* 14th ed., Automated Kjeldahl method, 7.021, Association of Official Analytical Chemists, Arlington, VA, 1984.
11. AOAC, *Official Methods of Analysis,* 14th ed., Fat in foods, 43.275, Association of Official Analytical Chemists, Arlington, VA, 1984.
12. AOAC, *Official Methods of Analysis,* 14th ed., Ash of animal feed, 7.009, Association of Official Analytical Chemists, Arlington, VA, 1984.
13. **Moore, S. and Stein, W. H.,** Procedures for chromatographic determination of amino acids on four percent cross-linked sulfonated polystyrene resins, *J. Biol. Chem.,* 211, 893, 1954.
14. **Schram, E., Moore, S., and Bigwood, E. J.,** Chromatographic determination of cystine as cysteic acid, *Biochem. J.,* 57, 33, 1954.
15. **Spies, J. R. and Chambers, D. C.,** Chemical determination of tryptophan, study of color-forming reaction of tryptophan, p-dimethyl-aminobenzaldehyde and sodium nitrite in sulfuric acid solution, *Anal. Chem.,* 20, 30, 1948.
16. **Spies, J. R.,** Determination of tryptophan in proteins, *Anal. Chem.,* 39, 1412, 1967.
17. AOAC, *Official Methods of Analysis,* 14th ed., Arsenic (total) in feeds, 42.035, Association of Official Analytical Chemists, Arlington, VA, 1984.
18. **Kakade, M. L., Rackis, J. J., McGhee, J. E., and Puski, G.,** Determination of trypsin inhibitor activity of soy products: a collaborative analysis of an improved procedure, *Cereal Chem.,* 51, 376, 1974.
19. **Tutelyan,.V. A., Eller, K. I., Sobolev, V. S., Sarafanov, A. G., Rubacova, N. V., and Urcumbaeva, T. N.,** *Manual on Detection and Content Determination of Aflatoxins, Deoxynivalenol (Vomitoxin) and Ochratoxin A in Agricultural Produce,* GKNT, Moscow, 1987, 36.

Chapter 5

A COMPARATIVE EXAMINATION OF SAFETY, BIOLOGICAL VALUE, AND DIGESTIBILITY

Vadim G. Vysotsky, Michael N. Volgarev, Irina S. Zilova, and Anatoli I. Zaichenko

The current concepts of how to increase food protein production are the results of scientific and technological developments in the agricultural and food processing industries. These developments have resulted in a policy to resolve the food protein availability issue on both short- and long-term bases for the food system. A primary focus for implementing this policy was to validate the potential of using nontraditional protein sources in the human diet. Potential new food protein sources from food processing waste and underutilized plant foods were identified. In addition, completely new protein food sources and new technologies were investigated. Isolated soy proteins and soy concentrates, containing 90 and 65% protein, respectively, showed the greatest potential for use as a raw material for the food industry. These processed protein sources also eliminate many of the antinutritional and organoleptically undesirable factors found in the starting materials.[1-5]

The first criteria which needs to be assessed with any new protein source is its safety for human consumption. The unusual origins of some protein sources make careful evaluations essential to ensure the health of the population. All of these new protein sources must be approached with the assumption that they may be carriers of toxic, allergenic, antinutritional, carcinogenic, and other undesirable components. In addition, the potential exists for the formation or addition of hazardous components during the processing of the new protein food sources.

Toxicity experiments must be conducted to determine the safety of new protein sources. These establish the safety of these new products through examination of critical organ tissues and blood analysis. Many known toxic substances can be identified by standard analysis and known safety levels. However, many products may contain unknown substances which can be detected only by feeding the products and measuring their effects through physiological changes in the animal. The safety evaluation should evaluate both biological hazards and chemical toxicants.

One of the primary objectives for studying the safety of protein products which are nontraditional in origin lies in revealing any possible general toxic effects and the nature of any primary damage to essential organs by carrying out experiments on laboratory animals. From the methodological standpoint the histopathological examinations are viewed as key observations, since biochemical and hematological changes are observed in the most cases only following a histological observation of the pathology.[6,7] Therefore, changes in the activity of enzymes in the blood serum may not be as sensitive as structural changes on the cellular level in terms of identifying toxicity. Correct assessment in changes of biochemical indices depends on how closely these data are correlated to morphological shifts.[6-10] Substrate and enzymological examinations of the blood serum are done in order to arrive at early diagnoses of possible damage to the visceral organs of animals ingesting new protein products, but cannot serve as reliable specific criteria for assessing their quality, since changes in the values for these indices are observed both when exposure to stress and after histopathological changes have occurred.[6-8]

The subchronic animal evaluations of the protein food is conducted only after the product has been thoroughly evaluated chemically and microbiologically. If the product is found to be free or within permissible levels of known toxic materials, then it is justified to determine whether or not unidentified safety hazards exist.

Another important aspect in studying the quality of a protein lies in determining the biological value and digestibility of the protein. These factors describe the degree to which the protein satisfies the amino acid requirements of human beings, but also the possible deterioration in the original aminogram for the native protein during extraction from raw material. Unfortunately, none of the existing methods for determining biological value gives a good enough description of its potential value for each protein for humans. This is because the existing methods for evaluating biological quality do not correlate with the new Food and Agriculture Organization/World Health Organization/United Nations University (FAO/WHO/UNU) pattern of amino acid requirements for 1985 employed during the calculation of the amino acid score.[11-13] The broad spectrum of biological methods is employed for determining biological value, with each method having its own merits and shortcomings, which makes it difficult to standardize methodology. In light of the above, calculating average biological values using various methods seems to be an alternative to these difficulties.

During the examinatin of new protein sources the safety, biological value, and digestibility were evaluated on isolated soya protein, Supro® 500E, and Supro® 710 made by Protein Technologies International (U.S.), and Ardex-M and Ardex-SP6 made by the ADM® Company (U.S.), as well as soya protein concentrates Unico® made by the Unilever Company (Holland) and VNIIZH made by the All-Union Scientific Research Institute of Fats (U.S.S.R.).

The safety of these isolated soya protein and soya concentrates was examined in a 3-month subchronic experiment. Growing male Wistar rats initially weighing 90 to 100 g were used. Two protein levels (18 and 30%) were used for the experimental (soya) or control (casein) proteins in semisynthetic rations as the sole sources of protein. Each experimental or control group consisted of 20 animals. Ten animals were decapitated 1 month into the experiment and the others were sacrificed at the conclusion of the study. The animals were weighed once a week over the course of the experiment. The absolute and relative weights of the liver, kidneys, heart, testicles, and spleen were determined following decapitation. These organs were further subjected to histological examination.

Chemical and biological methods were employed in carrying out the examinations on the biological value and digestibility of the isolated soya protein and soya concentrates. The amino acid score of proteins relative to the 1985 FAO/WHO/UNU reference amino acid pattern[13] was calculated as a measure of biological value. The biological evaluation involved 28 d study with growing male Wistar rats weighing an average of 50 to 55 g each and housed in separate cages. The content of experimental or control proteins in the rations as the sole source of protein nitrogen in the food was kept at a constant level (8 to 10%). The weight of the animals and the amount of food consumed were determined daily. The urine and feces of the animals as well as any remaining uneaten food were collected during the last 5 d and the total nitrogen content was determined using the Kjeldahl method.[11] The nitrogen data obtained were used for calculating the subsequent indices for biological value: protein efficiency ratio (PER), net protein ratio (NPR), true net protein digestibility (NPUtr), as well as its true digestibility (Dtr).

The calculation of the NPR, NPUtr, and digestibility include corrections for endogenous nitrogen losses based on previous studies with similar groups of male Wistar rats consuming nitrogen-free diet.s[14]

A summary of the body weight gain results from four experiments is given in Table 1. The body weight gain of the test and control animals did not differ statistically over the course of the experiment at the 18% protein content from isolated soya protein (Ardex-SP6 and Ardex-M) and concentrate (VNIIZH) in the ratio given to the rats. The only exception was the group receiving the Unico® soya protein concentrate. Similar results were obtained with the 30% protein diets. It is important to underscore that the use of the 30% protein level of any of the test or control proteins inhibited growth when compared to the 18% content of the same proteins in the rations, which confirms previously published literature.[15,16]

TABLE 1
Body Weight Gain of Rats Given Soya Protein
Isolates and Concentrates Over a Period of 90 d

Experiment number	Protein product	Protein content in rat diet, % of calories	
		18	30
1	Ardex-SP6	302.5 ± 24.2	269.9 ± 17.7
	Ardex-M	304.9 ± 15.4	283.7 ± 5.8
	Casein	324.2 ± 20.4	305.1 ± 25.0
2	Unico®	310.0 ± 19.6[a]	260.0 ± 11.0[a]
	VNIIZH	345.0 ± 18.3[a]	320.2 ± 10.1
	Casein	372.10 ± 15.5[a]	343.3 ± 18.6
3	PP500E	—	322.1 ± 17.4
	Casein	—	345.2 ± 19.2
4	PP710	—	312.9 ± 12.1
	Casein	—	337.8 ± 14.2

[a] Indicates significantly different at $p < 0.05$.

The body weight increased less in the test rats than in the control animals in each of the experiments conducted. This apparently was due primarily to the lower amount of sulfur-containing amino acids in the soya proteins in comparison to casein, which is corroborated by the data presented in Table 2.

No significant differences between the test and control animals were found while comparing the relative weight of several visceral organs (liver, kidneys, heart, testicles, and spleen) expressed in percent of body weight. Histological examinations of these organs failed to turn up any pathological changes with the exception of a slightly pronounced peritoneal small-drop fatty infiltration of the liver. The extent of this infiltration was more apparent in the test animals given an 18% protein diet from soya proteins than in the control animals and those given a 30% protein level of isolates and concentrates in the diets. This may be due to the differing extent to which these proteins and casein were limiting in the sulfur-containing amino acids.

The examinations therefore did not show that the soya proteins which were evaluated in these studies had any toxic effect on the laboratory animals. The observed low-level fatty infiltration of the liver was natural and could not be viewed as a barrier to using these protein products in food.

Another important aspect in evaluating the quality of isolated soya proteins and concentrates involved the determination of their biological value. The amino acid scores of the proteins are presented in Table 2 based on the amino acid analysis conducted in our laboratory. It is clear from the data presented in this table that all soya protein preparations studied are characterized by a different amino acid content than casein. The values for the amino acid score were somewhat lower for the Ardex-M and Ardex-SP6 proteins than in the other protein products examined. The amino acid scores of casein, isolated soy protein, PP500 and PP710, and concentrates Unico® and VNIIZH equaled 1.0 based on the FAO/WHO/UNU suggested pattern of requirement for the 2- to 5-year-old child.

The biological value of isolated soya proteins and concentrates was further examined in experiments with laboratory animals. The results of these examinations are presented in Table 3. The data in Table 3 indicate first and foremost the lower biological value with rats of the isolated soya protein and concentrates in comparison to casein regardless of the method used in its determination. This may easily be explained by the more pronounced limitation exerted by the sulfur-containing amino acids on the soya proteins in comparison to the casein

TABLE 2

Amino Acid Composition (A) and Amino Acid Score (C) of Isolated Soy Protein and Soy Concentrates

Essential amino acid	1985 FAO/WHO/UNO pattern A	Casein A	C	PP500E A	C	PP710 A	C	Ardex-SP6 A	C	Ardex-M A	C	Unico® A	C	VNIIZH A	C
Isoleucine	2.8	5.4	1.93	4.9	1.75	4.9	1.75	4.8	1.71	4.9	1.75	5.2	1.82	5.1	1.82
Leucine	6.6	9.5	1.43	8.2	1.24	8.2	1.24	8.1	1.23	8.3	1.26	7.6	1.15	7.8	1.18
Lysine	5.8	8.1	1.43	6.3	1.04	6.3	1.09	6.2	1.07	6.2	1.07	6.2	1.07	7.1	1.22
Methionine + cystine	2.5	3.2	1.28	2.6	1.04	2.6	1.04	2.7	1.08	2.7	1.08	2.8	1.12	2.6	1.04
Phenylalanine + tyrosine	6.3	11.1	1.76	9.0	1.43	9.0	1.43	8.5	1.35	8.5	1.35	8.7	1.38	8.5	1.35
Threonine	3.4	4.7	1.38	3.8	1.12	3.8	1.12	3.9	1.15	4.0	1.18	4.0	1.18	3.8	1.12
Tryptophan	1.1	1.4	1.27	1.4	1.27	1.4	1.27	1.0	0.91	1.0	0.91	1.5	1.36	1.6	1.45
Valine	3.5	7.7	2.20	5.0	1.43	5.0	1.43	5.0	1.43	5.3	1.51	5.1	1.46	5.2	1.49
Amino acid score	1.0		1.0		1.0		1.0		0.91		0.91		1.0		1.0	1.0

Note: A = Grams amino acid per 100 g protein. C = Amino acid score (grams amino acid per 100 g protein ÷ (grams acid req. per 100 g protein).

TABLE 3
Biological Value and Digestibility of Isolated Soy Proteins and Soy Concentrates

Exp. no.	Biological index	Casein	Isolated soy protein						Soy protein concentrates					
			PP500E		PP710		Ardex-SP6		Ardex-M		Unico®		VNIIZH	
		A	A	O	A	O	A	O	A	O	A	O	A	O
							10% Protein Diet							
I	PER	2.60					2.32	89	2.45	94	2.46	94	2.20	84
	NPR	3.26					2.95	90	3.11	95	3.14	96	2.81	86
	NPUtr	79.6					62.5	78	58.4	73	67.3	84	65.7	82
	Digestibility	95.1					95.1	100	94.4	99	91.5	96	91.1	96
							9.0% Protein Diet							
II	PER	2.94	2.43	83										
	NPU	3.44	3.31	96										
	NPUtr	78.3	68.4	87										
	Digestibility	97.9	95.9	98										
III	PER	3.26			2.83	87								
	NPR	3.96			3.69	93								
	NPUtr	59.1			56.0	95								
	Digestibility	96.0			95.3	99								

Note: (A) Absolute values; (O) value relative to casein, percent.

(Table 2). Of note as well is the lack of any common pattern with respect to the differences in the values for the relative biological value of the test proteins determined by the "growth" (PER, NPR) and "balanced" (NPU) methods. While the relative numbers for biological value calculated according to the results obtained from "growth" examinations were greater than those determined by means of the "balanced" method with respect to Ardex-SP6, Ardex-M, and Unico®, this trend was the reverse for the PP710 product. The protein preparations VNIIZH and PP500E had an intermediate position in this regard.

The above differences therefore made it impossible to calculate the relative biological value for the isolated soya protein and concentrates. Given this fact, simply averaging the values obtained from all methods turned out to be a way out of the situation. One may assume on the basis of this average that the relative biological value of the isolated soya protein and concentrates is as follows, relative to the casein: 89% for PP500E, 92% for PP170, 86% for Ardex-SP6, 87% for Ardex-M, 91% for Unico®, and 84% for VNIIZH. It follows from these data that the quality of isolated soya protein PP500e and PP710 and of soya protein concentrate Unico® was somewhat higher in comparison to the other preparations examined. It is important to emphasize that these values in the majority of cases were in agreement with the results obtained from calculating the relative values for the amino acid score (Table 2). The variation may be due to the inaccuracy of the analysis carried out on the amino acid composition of the proteins, or with limitations inherent to each chemical method.[11] The application of these methods may therefore only be justified when the objective is to reveal the limiting amino acids and making a rough determination of biological value. For this reason, biological experiments are necessary for making a practical assessment regarding the potential biological value of proteins, while conclusive judgments require examinations on human beings.

The examination of digestibility (Table 3) revealed several differences between isolated soya protein and concentrates. While it hardly differed from that of casein and measured an average of 99% in relative values in all of the isolated soya proteins examined, the relative true digestibility was somewhat lower at 96% in the soya concentrates. The lower values for digestibility found for the protein concentrates may be related to the high level of fiber content in the products.

The examinations conducted therefore revealed that the technology for obtaining these forms of protein products has an influence on their quality, with the indices being somewhat better for PP500E and PP710 protein isolates and the Unico® soya protein concentrate.

REFERENCES

1. **Tolstoguzov, V. B.,** *Synthetic Foodstuffs,* Nauka, Moscow, 1978.
2. *Toxicants Occurring Naturally in Foods,* 2nd ed., National Academy of Science, Washington, D.C., 1973.
3. **Pokrovsky, A. A.,** *Metabolic Aspects of Food Pharmacology and Toxicology,* Meditsina, Moscow, 1979.
4. **Tutelyan, A. A.,** Toxic substances in food and their danger to human health, *Vopr. Pitan.,* 6, 10, 1963.
5. Codex Alimentarius Commission, FAO/WHO Report of Second Session of the Codex Committee on Vegetable Proteins, Ottawa, I, 5, 1982; Codex Alimentarius Commission, 15th Session, Alinorm 83, 30, Rome, 1983.
6. Principles and Methods for Evaluating the Toxicity of Chemical Substances. Part I. Hygienic Criteria for the State of the Environment, World Health Organization, Geneva, 1981, 6.
7. **Bonashevskaya, T. I., Belyaeva, N. N., Kumpon, N. V., and Panasyuk, L. V.,** *Morphofunctional Examinations in Hygiene,* Meditsina, Moscow, 1984.
8. **Grice, H. C.,** The changing role of pathology in modern safety evaluation, in *CRC Crit. Rev. Toxicol.,* I, 119, 1972.
9. **Naji, A. A.,** Decreased activity of commonly measured serum enzymes: causes and clinical significance, *Am. J. Med. Technol.,* 49(4), 241, 1983.

10. **Adolf, U. and Lorenz, E.,** *Enzyme Diagnosis in Diseases of the Heart, Liver and Pancreas,* S. Karger, Basel, 1982.
11. **Pellet, P. L. and Young, V. R.,** *Nutrition Evaluation of Protein Foods,* UNU, Food and Nutrition Bulletin, Suppl. 4, 1980, 26.
12. **Pineda, O., Torun, B., Vitari, F. M., and Arroyave, G.,** Protein quality in relation to estimates of essential amino acid requirements, in *Protein Quality in Humans: Assessment and In Vitro Estimation,* AVI Publishing, Westport, CT, 1981, 3, 29.
13. *Energy and Protein Requirements,* Report of a Joint FAO/WHO/UNU Expert Consultation, Tech. Rep. No. 724, 1985.
14. **Vysotsky, V. G. and Mamaeva, E. M.,** Evaluating endogenic losses of nitrogen in white rats of various ages, *Vopr. Pitan.,* 3, 13, 1979.
15. **Edozien, Y. O. and Switze, B. R.,** Influence of diet on growth in the rat, *J. Nutr.,* 108(2), 262, 1978.
16. **Mamaeva, E. M.,** The influence of the content of casein in the rat food on the growth and condition of the nitrogen balance of white rats, *Vopr. Pitan.,* 1, 14, 1979.

Chapter 6

NUTRITIONAL VALUE OF SOYBEAN PROTEIN FOODS

Fred H. Steinke

The soybean as a food protein is widely available throughout the world. The production of soybeans has expanded greatly into temperate climate areas of both the northern and southern hemispheres, providing a large and reliable source of vegetable protein for use in the diet. The use of soybeans as a food was developed over many centuries in China. Many of these foods are highly acceptable to the Chinese and Japanese cultures, but less acceptable to nonoriental cultures. Therefore, new food products have been developed which can be used in Western-type foods.

The soybean production in the U.S. was developed originally as a source of vegetable oil. The protein residue was used primarily as a protein source to balance the protein and amino acid content of cereals for animal feeds. The use of soybean products has a 60-year history of animal feeding in the U.S. The use of soybean meal has also expanded through imports into Europe and the U.S.S.R. Thus, there is a long history of soybean use with very large numbers of animal species at all stages of life to validate its nutritional value and safety as a food protein source.

The wide availability and use with animal feeds resulted in a natural interest in using the protein source in human foods. Considerable technological effort has been committed, over the last 40 years, to developing soybean protein products which are acceptable and safe for use in foods. These products are now used in a wide range of foods as economical protein ingredients. In many cases, these soy protein foods replace other high-quality protein sources in traditional foods which are in short supply, too expensive, or undesirable for good health.

The soy products which are commonly available for use in food formulation are

Isolated soy protein
Soy protein concentrate
Soy flour, defatted
Soy flour, full fat
Whole soybeans

These products differ widely in their composition, functional characteristics, and acceptability by the consumer. The compositions of these products are given in Table 1 based on data published in the U.S. Department of Agriculture Agricultural Handbook[1] and from Protein Technologies International.[2] The isolated soy protein has the highest level of soy protein (92%) with only trace amounts of lipid and carbohydrates. Therefore, this product provides minimal effects of carbohydrates on flavor and acceptability. Soy protein concentrate contains 67.5% protein and 27% carbohydrate. This results in a product with lesser functional uses in the food industry. Defatted soy flour contains 52.6% protein and 35% carbohydrates, which further reduces its value as a protein source and contribution as a functional ingredient. Full fat soy flour contains only 40% protein with 23% lipid and 32% carbohydrates. The high lipid content adds to the caloric contribution of this product. The high polyunsaturated fat content can stimulate increased oxidization.

The trace element content of isolated soy protein is given in Table 2. The isolated soy protein contains low levels of most elements. Isolated soy protein is a significant source of iron with 140 ppm. The sodium and potassium levels depend to a large extent on the cation

TABLE 1
Composition of Soybean Protein Products — Dry Basis

Soybean protein product	Protein (N × 6.25) (g/100 g)	Lipid (g/100 g)	Carbohydrate (g/100 g)	Ash (g/100 g)	Energy (kcal/100 g)	Ref.
Isolated soy protein	92.0	1.1	2.6	4.3	388	2
Soy protein conc	67.5	0.5	27.0	5.0	348	1
Soy flour, defatted	52.6	1.2	34.7	6.3	335	1
Soy flour, full fat	39.6	22.7	31.6	6.1	456	1
Whole soybeans	43.7	21.8	29.2	5.3	455	1

TABLE 2
Typical Compositions of Isolated Soy Protein PP500E and PP710

Element		Content PP710	Content PP500E
Calcium	%	0.16	0.16
Phosphorus	%	0.77	0.74
Potassium	%	0.084	0.083
Sodium	%	1.3	1.1
Cobalt	ppm	2.0	2.0
Copper	ppm	15	15
Iron	ppm	140	140
Manganese	ppm	14	14
Molybdenum	ppm	3.0	3.0
Selenium	ppm	0.11	0.10
Zinc	ppm	34	34
Heavy metals			
Arsenic	ppm	<2	<0.2
Cadmium	ppm	<0.2	<0.2
Lead	ppm	<0.2	<0.2
Mercury	ppm	<0.05	<0.05

Summarized from Technical Data Sheets PP-500E and PP-710, Protein Technologies International, St. Louis, MO.

used to neutralize the protein during processing. If the proteinate is neutralized with potassium rather than sodium, the potassium level can increase to 1.6 to 1.8% and the sodium will be reduced to less than 0.1%. Heavy metals are low in these products and do not pose a safety concern. In addition, these products are carefully monitored during processing to ensure that they remain within safety specifications.

Soybean proteins contain all of the 18 amino acids found in typical food proteins (Table 3). These include all nine of the essential amino acids, as well as the semiessential and nonessential amino acids. The essential amino acid content in milligrams per gram of isolated soy protein are presented in Table 4. In addition, the amino acid requirements of the preschool child, school child, and adult[3] and the growing rat[4] are presented in this table for comparison. The amino acid content of isolated soy protein exceeds the suggested pattern of requirement for the preschool child, school child, and adult human based on these new patterns published by the World Health Organization in 1985.

The growing rat is a typical animal model which is used to evaluate protein sources.

TABLE 3
Amino Acid Composition of Typical Isolated Soy Protein

Amino acid	Grams amino acid per 100 g product	Grams amino acid per 100 g protein
Essential amino acids		
Histidine	2.3	2.6
Isoleucine	4.3	4.9
Leucine	7.2	8.2
Lysine	5.5	6.3
Methionine	1.2	1.3
Cysteine	1.1	1.3
Total sulfur amino acids	2.3	2.6
Phenylalanine	4.6	5.2
Tyrosine	3.3	3.8
Total aromatic amino acids	7.9	9.0
Threonine	3.3	3.8
Tryptophan	1.1	1.3
Valine	4.4	5.0
Nonessential amino acids		
Alanine	3.8	4.3
Arginine	6.7	7.6
Aspartic acid	10.2	11.6
Glutamic acid	16.8	19.1
Glycine	3.7	4.2
Proline	4.5	5.1
Serine	4.6	5.2

Adapted from Technical Data Sheet PP-500E, Protein Technologies International, St. Louis, MO.

TABLE 4
Essential Amino Acid Content of Isolated Soy Protein and Patterns of Requirement for Humans (milligrams amino acid per gram of protein)

Essential amino acid	Isolated soy protein[2]	Suggested pattern of requirement[3]			Growing rat
		Preschool child	School child	Adult	
Histidine	26	19	19	16	25
Isoleucine	49	28	28	13	42
Leucine	82	66	44	19	62
Lysine	63	58		16	58
Total sulfur amino acid	26	25	22	19	50
Total aromatic amino acid	90	63	22	19	67
Threonine	38	34	28	9	42
Tryptophan	14	11	9	5	125
Valine	50	35	25	13	50
Arginine	76	—	—	—	50

However, the growing rat has a quite different amino acid requirement than does the human. The growing rat has a higher requirement for histidine, isoleucine, total sulfur amino acids (methionine and cystine), threonine, and valine. In addition, the growing rat requires arginine, which is not required by humans. The sulfur amino acid requirement of the rat is twice that of the preschool-age child. This high-sulfur amino acid requirement is the result of the high percentage of body hair of the rat which contains large amounts of sulfur amino acids. This high requirement for sulfur amino acid by the rat will result in a reduction in growth with protein sources that do not contain high levels of sulfur amino acid. These

TABLE 5
Protein Efficiency Ratio with the
Growing Rat with PP710

Test number	Adjusted PER
1	1.7
2	1.7
3	1.8
4	1.9
5	1.9
6	1.8
Mean	1.8

proteins may be completely adequate for the human who has a lesser need for sulfur amino acids. Legumes, in particular, may be adequate for humans while being inadequate for the growing rat. In any evaluation of protein, the human requirements should be the criteria of measurement. This has been recently recognized by the recommendation of the *digestibility corrected amino acid score* as the most appropriate method for evaluating protein quality in foods for human consumption.[5] This method of protein quality measurement uses the essential amino acid pattern of the 2- to 5-year-old child[3] as the reference pattern for amino acid scoring.

The biological values of isolated soy proteins have been evaluated in numerous studies to confirm their efficacy and safety as a protein source for human foods. The rat biological evaluation such as the protein efficiency ration (PER) is one such model which has been widely used, as has the net protein utilization (NPU). These types of studies can give some measure of the value of a protein. However, the interpretation of the results must take into account the differences in amino acid requirements between the rat and the human. Six PER assays conducted with PP710 (an isolated soy protein) are summarized in Table 5. The PER values range from 1.7 to 1.9, with a mean value of 1.8 relative to casein at a PER of 2.5. The values are similar to those obtained with another isolated soy protein, PP500E (Supro®-620), which had an average PER of 1.82 ± 0.15 based on 24 assays.[6] The values are to be expected based on the sulfur amino acid content of isolated soybean protein and casein. The sulfur amino acid content of isolated soybean protein is 2.6 g/100 g protein, and casein is 3.5 g/100 g protein.[6] The sulfur amino acid content of isolated soy protein is therefore 74% of that of casein. The PER values of isolated soy proteins based on the results cited above are 72% of casein (1.8 − 2.5) × 100. This means that the difference in PER values is due solely to the difference in sulfur amino acid content in these two protein sources. The amino acids must therefore be fully available and free of negative factors which could interfere with the amino acid utilization.

Similar results have been obtained with the young pig (14 to 40 d of age) using NPU measurements. The results of three experiments comparing PP710 with either skim milk protein or casein are summarized in Table 6. NPU values ranged from 73 to 77.7 with PP710, while NPU values for milk proteins ranged from 87 to 89. The value of PP710 relative to the milk protein was 83.2 to 87.3. This value is higher than that obtained with the PER using the rat model. This is also to be expected since the sulfur amino acid requirements of the baby pig[7] are less than that of the rat, but higher than that of the human.

The safety of the isolated soy protein has been confirmed by evaluations using the short term with the rat and direct human feeding studies. In addition, the use of isolated soy protein in human foods such as soy protein infant formulas for over 25 years confirms their safety and adequacy as a food-grade protein source. In the U.S. the isolated soy protein infant formulas now account for 25% of all infant formula sales.

In order to further document the adequacy and safety of the isolated soy protein, a

TABLE 6
Biological Evaluation of Isolated Soy Proteins with
Animal Models — NPU with the Baby Pig

Experiment no.	Skim milk	PP710	Percent of milk protein
1	87.7	73.0	83.2
2	87.2	76.1	87.3
Mean	87.5	74.6	85.2

	Casein	PP710	Percent of milk protein
3	89.1	77.7	87.2

TABLE 7
28-d Evaluation with the Growing Rat

	Percent protein in the diet			
Casein	30	20	10	0
PP710	0	10	20	30
Total protein	30	30	30	30
Weight gain[a]	215[b]	219[b]	225[b]	204[b]
Food consumption	442[b]	460[b]	470[b]	447[b]
g/Food/g gain	2.06[b]	2.11[ab]	2.09[ab]	219[a]
Liver weight, %[b]	5.31[a]	5.34[a]	5.23[ab]	4.87[b]
Kidney weight, %[b]	0.94[b]	0.92[b]	0.90[b]	0.88[b]

[a] Values in the same line not followed by the same letter are significantly different at $p < 0.05$.
[b] Presented as percent of body weight.

28-d study was designed with a ration containing 30% protein from casein. Isolated soy protein replaced one third, two thirds, or all of the protein in the diet to test whether or not the changes could be observed in the rats consuming those diets (Table 7). Weight gains and food consumption were similar for all four diet treatments, indicating no difference in the value of the protein sources and the absence of negative factors. Feed utilization was slightly lower only with the highest isolated soy protein diet. No significant differences were observed in kidney weight or in the histology of the organs taken at necropsy. Liver weight was significantly lower with the highest level of isolated soy protein. This later observation is considered a positive observation. Similar observations have been made with rats in long-term feeding studies, indicating an enlargement of kidneys with casein, but not with isolated soy protein.[8]

A direct evaluation of the acceptability of PP710 was conducted with human volunteers.[9] Two groups of 50 volunteers consumed either 40 g of PP710 or dried skim milk for 60 d, in addition to their normal diets. Clinical chemistries, hematology, and urinalysis were made initially and after 30 and 60 d on consuming the ISP.

All volunteers completed the study without any indication of physical changes. Gastrointestinal functions were normal with no indication of increased gas, nausea, loose stools, or constipation. None of the subjects on either skin milk control or ISP experienced any allergic responses.

The results of the initial and the 60-d clinical chemistries, hematology, and urinalysis are summarized in Table 8. The overall results of these 20 clinical chemistries, 12 hematological, and 2 urinalyses demonstrate either normal values and minimal changes over the

TABLE 8

Average Clinical Chemistries, Hematology, and Urinalysis of Subjects Consuming 40 g/d of PP710 (PP) or Skim Milk (SM) Protein for 60 d

Blood Chemistry	Initial	60 d	Change
Albumin (g/100 ml)			
SM	4.36 ± 0.49	4.65 ± 0.28	0.29
PP	4.38 ± 0.39	4.63 ± 0.29	0.25
Alkaline phosphatase (IU/l)			
SM	42.12 ± 12.65	41.62 ± 12.61	−0.50
PP	44.56 ± 16.35	44.96 ± 16.81	0.40
Uric acid (mg/100 ml)			
SM	4.83 ± 1.24	5.26 ± 1.21	0.43
PP	4.93 ± 1.34	5.63 ± 1.32	0.70
Chloride (meq/l)			
SM	101.74 ± 2.12	100.94 ± 2.64	−0.80
PP	101.56 ± 2.35	100.72 ± 2.46	−0.84
Glucose (mg/100 ml)			
SM	87.30 ± 7.21	82.28 ± 8.38	−5.02
PP	87.78 ± 8.39	81.96 − 5.93	−5.82
Blood urea nitrogen (mg/100 ml)			
SM	14.18 ± 2.95	14.32 ± 3.53	0.14
PP	13.34 ± 3.34	13.38 ± 3.87	0.04
Sodium (meq/l)			
SM	142.22 ± 2.36	137.18 ± 3.18	−5.04
PP	141.60 ± 3.36	138.58 ± 5.78	−3.02
Potassium (meq/l)			
SM	4.11 ± 0.28	4.00 ± 0.31	−0.11
PP	4.10 ± 0.27	3.98 ± 0.26	0.12
Calcium (mg/100 ml)			
SM	9.55 ± 0.53	9.92 ± 0.47	0.37
PP	9.61 ± 0.45	9.95 ± 0.46	0.34
Phosphorus (mg/100 ml)			
SM	4.34 ± 0.51	3.98 ± 0.56	0.35
PP	4.15 ± 0.83	4.02 ± 0.52	−0.12
Creatinine (mg/100 ml)			
SM	1.03 ± 0.11	1.20 ± 0.20	0.16
PP	1.07 ± 0.19	1.17 ± 0.19	0.10
Total protein (g/100 ml)			
SM	7.19 ± 0.41	7.28 ± 0.35	0.10
PP	7.16 ± 0.43	7.28 ± 0.36	0.12
Bilirubin (total mg/100 ml)			
SM	0.57 ± 0.37	0.71 ± 0.59	0.14
PP	0.71 ± 0.51	0.65 ± 0.36	−0.06
Bilirubin, direct (mg/100 ml)			
SM	0.16 ± 0.09	0.19 ± 0.13	0.03
PP	0.18 ± 0.09	0.19 ± 0.08	0.01
Lactate dehydrogenase (IU/l)			
SM	71.54 ± 14.06	72.36 ± 12.95	0.82
PP	74.58 ± 15.00	79.94 ± 43.80	5.36
Amylase (units)			
SM	121.76 ± 25.70	111.74 ± 26.10	−10.02
PP	125.82 ± 29.96	111.34 ± 25.75	−14.48
SGOT (IU/l)			
SM	13.72 ± 4.76	14.14 ± 3.93	0.42
PP	13.62 ± 4.26	19.78 ± 41.33	6.16
SGPT (IU/l)			
SM	11.58 ± 4.54	15.12 ± 5.98	3.54
PP	11.02 ± 4.58	15.96 ± 12.11	4.94
Total cholesterol (mg/100 ml)			

TABLE 8 (continued)
Average Clinical Chemistries, Hematology, and Urinalysis of Subjects Consuming 40 g/d of PP710 (PP) or Skim Milk (SM) Protein for 60 d

Blood Chemistry	Initial	60 d	Change
SM	185.44 ± 47.43	193.72 ± 42.26	8.28
PP	194.24 ± 41.91	199.08 ± 44.52	4.84
Triglycerides (mg/100 ml)			
SM	94.42 ± 42.73	88.42 ± 34.29	−6.00
PP	101.16 ± 72.23	102.26 ± 74.82	1.10

Hematology

	Initial	60 d	Change
Hemoglobin (g/100 ml)			
SM	13.85 ± 1.31	13.41 ± 1.30	−0.43
PP	13.77 ± 1.41	13.38 ± 1.32	−0.40
Hematocrit			
SM	41.20 ± 3.43	40.50 ± 3.76	−0.70
PP	41.12 ± 3.77	40.54 ± 3.67	−0.58
White blood cells × 10^3/100 ml			
SM	6.57 ± 1.72	6.62 ± 2.14	0.04
PP	6.98 ± 1.94	6.82 ± 1.61	−0.16
Segmented cells (%)			
SM	53.46 ± 8.14	53.12 ± 7.39	−0.34
PP	56.12 ± 7.68	57.96 ± 8.06	1.84
Bands (%)			
SM	0.08 ± 0.27	0.20 ± 0.90	0.12
PP	0.10 ± 0.46	0.12 ± 0.59	0.02
Lymphocytes (%)			
SM	35.86 ± 8.36	35.44 ± 7.69	−0.42
PP	32.94 ± 7.27	30.78 ± 7.14	−2.16
Monocytes (%)			
SM	7.12 ± 2.80	7.98 ± 2.83	0.86
PP	7.08 ± 2.84	7.30 ± 3.53	0.22
Eosinophils (%)			
SM	2.70 ± 2.52	2.68 ± 2.13	−0.02
PP	3.20 ± 3.26	3.30 ± 2.79	0.10
Basophils (%)			
SM	0.26 ± 0.56	0.26 ± 49	0.00
PP	0.18 ± 0.44	0.20 ± 0.53	0.02
Atypical lymphocytes (%)			
SM	0.56 ± 1.23	0.34 ± 0.87	−0.22
PP	0.40 ± 0.99	0.16 ± 0.51	−0.24
Prothrombin time (s)			
SM	14.48 ± 0.80	13.63 ± 0.97	−0.85
PP	14.44 ± 0.98	13.71 ± 1.03	−0.74

Urine Chemistries

	Initial	60 d	Change
Urine specific gravity			
SM	1.02 ± 0.01	1.03 ± 0.01	0.01
PP	1.02 ± 0.01	1.03 ± 0.01	0.01
Urine pH			
SM	5.60 ± 0.56	5.37 ± 0.59	−0.23
PP	5.58 ± 0.77	5.52 ± 0.65	−0.06

60-d feeding period for either the skim milk or the PP710. These data confirm the long experience with isolated soy protein as a safe and acceptable protein source for the human.

Soy protein foods and ingredients have been used as food protein sources for many years with a history of safe and acceptable use. The long history is supported by analytical and biological data demonstrating adequacy of the amino acid content for human requirements. Both animal evaluations and human feeding studies support the safety of these protein sources for the human food system.

REFERENCES

1. Composition of Foods: Legume and Legume Products, Agricultural Handbook 8-16, Revised December 1986, U.S. Department of Agriculture, Washington, D.C.
2. Technical Data Sheet PP-500E, Protein Technologies International, St. Louis, MO.
3. Energy and Protein Requirements, WHO Tech. Rep. Ser. 724, World Health Organization, Geneva, 1985.
4. Nutrient Requirements of the Laboratory Rat, 3rd ed., National Research Council, National Academy of Sciences, Washington, D.C., 1978.
5. Protein Quality Evaluation, Report of a Joint FAO/WHO Expert Consultation, Food and Agriculture Organization, Rome, 1990.
6. **Steinke, F. H., Prescher, E. E., and Hopkins, D. T.,** Nutritional evaluation (PER) of isolated soybean protein and combinations of food proteins, *J. Food Sci.,* 45, 323, 1980.
7. Nutrient Requirements of Swine, 8th ed., National Research Council, National Academy of Sciences, Washington, D.C., 1979.
8. **Iwasaki, R., Gleiser, C. A., Masoro, E. J., McMahan, C. A., Seo, E. J., and Yu, B. S.,** The influence of dietary protein source on longevity and age-related disease processes of fischer rats, *J. Gerontol.,* 43, 1, B 5-12, 1988.
9. **Scrimshaw, N. S., Young, V. R., and Steinke, F. H.,** The Tolerance of a Soy Protein Isolated in Adult Humans, Department of Nutrition and Food Science, Massachusetts Institute of Technology, Cambridge, unpublished data.

Chapter 7

THE BIOLOGICAL VALUE AND DIGESTIBILITY OF PROTEINS IN COMBINED FOOD PRODUCTS MADE WITH ISOLATED SOY PROTEIN OR SOY CONCENTRATES

Vadim G. Vysotsky, Michael N. Volgarev, and Anna M. Safronova

Success in the incorporation of new and additional sources of protein in human nutrition hinges primarily on the following three fundamental conditions:

1. Finding manufacturing conditions which ensure that the end protein products are completely safe
2. The comprehensive study of the chemical composition and biological efficiency of new protein products
3. The development of a scientific basis for the optimal utilization of these proteins for nutritional goals

After new protein products have been proven safe, the primary focus shifts to knowledge about their biological value and above all about the amino acid composition not only of these individual proteins, but of the total food proteins as the basis for building balanced protein compositions in the diet.

Human beings usually consume a mixture of proteins from products of animal and vegetable origin. The natural inclination on the part of humans to diversify the foods consumed is due both to a desire to improve its taste and the physiological/biochemical mechanisms for maintaining homeostasis, which depends on the amino acid balance in the total quantity of proteins. Diets remain limited in many areas of the globe due primarily to a shortage of animal proteins. In some special cases such as enteral diets and infant formulas, products are single-protein diets based on soya or cow milk proteins.

The majority of proteins,[1,2] including soya protein and milk protein, are limited with respect to individual essential amino acids when their aminograms are compared to the 1973 Food and Agriculture Organization/World Health Organization (FAO/WHO) reference profile of amino acids.[3] In 1985, the FAO/WHO/United Nations University (UNU) came out with a new amino acid pattern,[4] in which the soya proteins and milk proteins were not limiting in essential amino acids. Additional evidence showing the essentiality of amino acid profile is required to validate their use in evaluating food proteins. Studies are currently in progress aimed at finding out whether or not highly valuable proteins may in certain types of foods be replaced with ones which are less expensive, but which in a majority of cases are not of as high a quality. This has the objective of saving on the amount of animal proteins, increasing the utilization of additional sources of protein and raising the biological value of vegetable proteins.

Experience has shown that simply introducing protein isolates or concentrates in traditional foodstuffs with the objective of increasing the overall content of protein or partially replacing it based solely on consideration of the production process and on consumer characteristics frequently fails to effect the desired increase in the biological value of the proteins in the enriched product. It also does not ensure that its original level will be maintained. A similar situation is encountered during the development of infant formulas and enteral products from a single protein base, since the amino acid composition of this source generally does not correspond to the amino acid needs of the recipient. It is also a well-established fact that animals best utilize protein when given a balanced amino acid composition not just in daily rations, but each time food is ingested.[5-8]

The utilization of food increases when complete proteins in terms of essential amino acids are achieved by using a mixture of animal and vegetable proteins or of animal proteins exclusively. One of the most important jobs therefore becomes one of developing combined products which are inexpensive and accessible, while at the same time have a higher biological value or efficiently replace highly valuable proteins with lower-quality proteins without reducing the biological value of the final products. The use of new sources of food protein may be of utmost importance in this regard. It is clear that the development of protein compositions based on the effects of mutual enrichment or substitution[6,9-11] as set forth in theories relating to the concept of amino acid balance[7] and the feasibility for its realization represents the most efficient way to go about using new resources of protein for nutritional objectives.

Information has been developed concerning the quantitative requirements of human beings for essential amino acids as well as the most adequate ratio for a reference amino acid pattern[3,4,12] based on the concept of amino acid balance. This concept assumes that the maximum efficiency with which the protein may be utilized by the animal organism depends on how close the proportions of amino acid in this protein come to the proportions of the amino acids actually required by human beings after a correction for digestibility.[7] In order to achieve mutual protein enrichment or justify protein substitution, one must therefore take into account principles based upon the amino acid balance concept.

The principle of differentiation also applies with regard to the requirements for essential amino acids in various segments of the population, the specificity of the amino acid composition of food proteins, as well as the digestibility. The development of protein compositions with the aim of retaining or increasing the biological value of proteins in traditional foodstuffs or as balanced foundation for special types of foods (for infants, therapy, and prevention of chronic diseases) with the use of new sources should therefore not only be based on technological principles, but avail itself of biomedical concepts related to amino acid balance and imbalance as well.

The process of developing balanced protein compositions or protein products higher in biological value should consist of three stages: planning the product based on calculations of the amino acid balance, manufacturing the product with the set chemical characteristics from the calculation, and subjecting the product to a biomedical examination for purposes of quality evaluation. In an effort to find out the advisability of this sequence of stages for developing new combined protein products, studies were carried out using sausage products ("Stolovaya", boiled sausage) and bread.

In the first stage of these examinations, the amino acid composition of the primary initial protein components included in the recipe of both traditional and combined types of the model sausage and bread were studied. The results obtained from these examinations were used to calculate values for the amino acid score for each protein.[3,12] The data obtained served as the basis for the computer-assisted theoretical determination of the amino acid profile and score given varying degrees of protein substitution in the meat (beef/pork blend) by the PP500E isolated soya protein, as well as for establishing the optimum proportion of wheat flour protein and the PP500E soya protein isolate at which the maximum possible effect of mutual enrichment is achieved.

Experimental samples of test (combined) and control sausage products and breads were prepared by institutes of the food industry in the second step of the evaluation. The meat proteins in the combined sausage products were replaced by soya protein isolates at 12.5, 25.0, 50.0, and 100.0%. Only one portion of wheat flour and soya protein isolate was employed during the development of the combined bread product, since the primary objective in this examination was to achieve a maximum level of mutual protein enrichment. According to results from computer calculations, this manifested itself at a ratio of wheat flour to soya protein isolate equal to 79:21 on a product basis or 29:71 on a protein basis.

TABLE 1

The Amino Acid Content (A) and Amino Acid Score (C) of Protein Products Used to Develop Blended Food Products

Essential amino acids	FAO/WHO/UNU 1985 pattern A	Beef A	Beef C	Pork A	Pork C	Wheat flour 1st grade A	Wheat flour 1st grade C	Isolated soy protein A	Isolated soy protein C
Isoleucine	2.8	4.2	1.50	4.9	1.75	3.7	1.32	4.9	1.75
Leucine	6.6	8.0	1.21	7.5	1.14	7.0	1.06	8.2	1.24
Lysine	5.8	8.5	1.46	8.7	1.50	2.1	0.36	6.3	1.09
Methionine + cystine	2.5	3.8	1.52	3.7	1.48	4.0	1.60	2.6	1.04
Phenylalanine + tyrosine	6.3	7.8	1.24	7.7	1.22	7.2	1.14	9.0	1.43
Threonine	3.4	4.3	1.26	4.6	1.35	2.7	0.79	3.8	1.11
Tryptophan	1.1	1.1	1.00	1.3	1.18	1.1	1.00	1.4	1.27
Valine	3.5	5.6	1.60	5.8	1.66	4.1	1.17	5.0	1.43

Note: A = Grams of amino acid per 100 g proteins; C = amino acid score (g/100 g protein ÷ g requirement per 100 g protein).

The third stage of the examination involved biomedical studies of the biological value and digestibility of the proteins in the combined products. Biological value was assessed by calculating the amino acid score[4,12] as well as in experiments on laboratory animals,[12] while digestibility was ascertained in experiments with laboratory animals.[12]

The results obtained from determining the amino acid composition and calculating the amino acid score relative to the FAO/WHO/UNU 1985 reference amino acid pattern[4] for all protein-containing components in both types of combined foodstuffs are presented in Table 1.

As may be seen from the data presented, the proteins of the products listed in the table differ considerably in terms of amino acid composition when compared to each other and against the reference aminogram. Meat proteins and isolated soya protein are characterized by an "excess" amount of all essential amino acids in comparison to the reference amino acid pattern. The proteins in the wheat flour were limiting in several amino acids, both when compared to the reference pattern and against the amino acid composition of beef, pork, and soya proteins. The wheat flour proteins were limited in particular in terms of lysine and threonine. The highest content of sulfur-containing amino acids was found in the wheat flour protein, while the quantity of aromatic amino acids and tryptophan was greatest in the isolated soya protein. The lack of any amino acid limitation in certain proteins and the different amounts in others, therefore, makes it possible to select those protein compositions which are the most optimum. One may assume that the use of isolated soya protein during the production of combined meat and bread products should be based on the substitution effect, since a higher content of sulfur-containing amino acids in meat protein makes it possible up to a certain point to dilute them with soya protein having a lower content of these amino acids. The limiting factor during the development of combined bread products will be lysine.

In an attempt to test the above hypothesis, the values for the amino acid score of proteins in the combined sausage products were calculated with various amounts (0, 1.25, 25, 50, and 100%) of meat protein (ratio of beef and pork proteins: 52:48) substituted by isolated soya protein. Similar values were calculated for the amino acid score of proteins in the combined bread product given a ratio of wheat flour proteins to isolated soya protein, which was most optimal from a technological point of view. The results obtained from these calculations are presented in Table 2, and indicate that any amount up to 100% of the 52.48 beef and pork protein mixture normally employed during the production of boiled sausage

TABLE 2
Amino Acid Composition (A) and Amino Acid Score (C) of Protein Mixtures

| Essential amino acid | Meat product (percent isolated soy protein replacing meat protein) | | | | | | | | | | Bread flour (ratio of wheat: ISP protein) | | | | | |
| | 0 | | 1.25 | | 25 | | 50 | | 100 | | 100:0 | | 29:71 | | 11:91 | |
	A	C	A	C	A	C	A	C	A	C	A	C	A	C	A	C
Isoleucine	4.5	1.61	4.6	1.64	4.6	1.64	4.7	1.68	4.9	1.75	3.7	1.32	4.6	1.64	4.7	1.68
Leucine	7.8	1.18	7.8	1.18	7.8	1.18	8.0	1.21	8.2	1.24	7.0	1.06	7.8	1.18	8.0	1.21
Lysine	8.6	1.48	8.3	1.43	8.0	1.39	7.4	1.27	6.3	1.08	2.1	0.36	5.1	0.88	5.8	1.00
Methionine + cystine	3.8	1.52	3.6	1.44	3.4	1.36	3.2	1.28	2.6	1.04	4.0	1.60	3.0	1.20	2.8	1.12
Phenylalanine + tyrosine	7.8	1.24	7.9	1.25	8.0	1.27	8.4	1.33	9.0	1.43	7.2	1.14	8.5	1.35	8.8	1.40
Threonine	4.4	1.29	4.3	1.26	4.2	1.23	4.1	1.20	3.8	1.12	2.7	0.79	3.5	1.03	3.7	1.09
Tryptophan	1.2	1.09	1.2	1.09	1.2	1.18	1.3	1.18	1.4	1.27	1.1	1.00	1.3	1.18	1.4	1.27
Valine	5.7	1.63	5.6	1.60	5.5	1.57	5.3	1.51	5.0	1.43	4.1	1.17	4.3	1.37	4.9	1.40

Note: A = grams of amino acid per 100 g proteins; C = amino acid score (g/100 g protein + g requirement per 100 g protein).

TABLE 3

Biological Value and True Digestibility of Proteins in Blended Food Products

Measurement (%)	Meat product (percent isolated soy protein replacing meat protein)					Bread (ratio of wheat protein:protein)	
	0	12.5	25	50	100	100:0	29:71
Biological, value[a]	65.8 ± 1.1	65.1 ± 3.6	63.7 ± 3.6	59.1 ± 2.8[b]	57.7 ± 0.6[b]	41.7 ± 2.7	57.5 ± 3.4[b]
Digestibility, true	94.5 ± 0.7	95.0 ± 0.5	95.8 ± 0.4	94.2 ± 1.1	95.1 ± 0.4	85.4 ± 0.4	94.1 ± 0.8[b]

[a] Net protein utilization, true.
[b] Significant difference ($p < 0.05$).

products may be replaced by isolated soya proteins without in any way reducing the biological value of the protein in this mixture relative to the FAO/WHO/UNU 1985 reference pattern. Two ratios were calculated for the second combination (wheat flour + isolated soya protein). The first (protein ratio 11:89) yielded a score value equal to 1.0 (by lysine), while the score value was 0.88 (by lysine) in the second ratio (29:71). The second ratio was developed due to the fact that satisfactory bread could not be obtained with the 11:89 ratios. This condition could be satisfied by the second protein ratio, as well as lower proportions of isolated soya proteins. In light of the above, a second protein ratio equaling 0.88 (by lysine), as opposed to the 0.36 in wheat flour protein (by lysine), was selected for the examinations. The biological value of the proteins in the combined bread product calculated by method of amino acid score, therefore, rose 2.4 times relative to that of the unenriched product.

This method for determining biological value only provides a rough idea about its value, since it does not take into account the overall balance exhibited by the amino acid composition of the proteins, the difference in their digestibility, and the degree of biological availability of the individual amino acids. In light of the above, studies were carried out to determine the biological value inherent in the proteins of the traditional and combined products with rats. Table 3 presents the results obtained from these evaluations, which were carried out with the help of the "carcass" method.[12]

These evaluations established a constant reduction in the numbers for the biological value of overall protein in combined sausage products as the replacement of meat proteins with soya proteins increased. At 12.5 and 25% substitutions, the biological value did not differ statistically from the values for the control (no substitution). A further rise in the degree of substitution resulted in a significant ($p < 0.05$) drop in biological value. The data obtained indicate that meat proteins have a better amino acid balance than soya proteins for the growing rat. This diffrence is primarily due to the higher sulfur amino acid requirements of the rat vs. that of the human. This difference in requirement must be taken into account when evaluating these results.

A comparative determination of the biological value of proteins in traditional bread and bread enriched with a soya protein isolate revealed a pronounced effect. The biological values rose by a factor of 1.4 in the experimental bread sample in comparison to the conventional bread, which in principal corroborated the results obtained from calculating the amino acid score (Table 2).

The examination of protein digestibility showed virtually no differences between the control and experimental samples of sausage products. At the same time, the digestibility of the proteins in the combined bread product was significantly higher than that of the traditional bread ($p < 0.05$), which corroborates data in published literature.[13]

It should be emphasized in conclusion that using isolated soya proteins with the objective of replacing highly valuable animal proteins and enriching vegetable proteins of low value is justified from a biomedical point of view, which the results of the studies confirm. The data obtained also confirm the prognostic importance of the amino acid score method for carrying out tentative calculations of protein ingredients during the development of combined products as well as the necessity of carrying out biological experiments for determining the biological value of proteins.

REFERENCES

1. **Chernikov, M. P.**, *Proteolysis and Biological Value of Proteins,* Meditsina, Moscow, 1975, 171.
2. Amino Acid Content of Foods and Biological Data on Proteins, Food and Agriculture Organization, 1981.
3. Energy and Protein Requirements, Report of a Joint· FAO/WHO/UNU Expert Consultation, Tech. Rep. Ser. No. 522, World Health Organization, Geneva, 1973.
4. Energy and Protein Requirements, Report of a Joint· FAO/WHO/UNU Expert Consultation, Tech. Rep. Ser. No. 724, World Health Organization, Geneva, 1985.
5. **Pokrovsky, A. A.**, The determination of human nutritional requirements, *Vest. an SSSR,* 5, 3, 1964.
6. **Pokrovsky, A. A.**, Biochemical basis for developing products of elevated biological value, *Vopr. Pitan.,* 1, 5, 1964.
7. **Harper, A. E.**, Amino acid toxicities and imbalances, in *Mammalian Protein Metabolism,* II, Munro, H. N. and Allison, Y. B., Eds., Academic Press, New York, 1964, 87.
8. **Allison, Y. B.**, Biological evaluation of proteins, *Physiol. Rev.,* 35, 3, 664, 1955.
9. **Sharpenak, A. E. and Eremin, G. N.**, The possibility of mutual protein enrichment during the uniform ingestion of the latter with the food, *Vopr. Pitan.,* 4, 9, 1956.
10. **Bressani, R., Elias, L. G., and Gomez, B. R. A.**, Improvement of protein quality by amino acid and protein supplementation, in *Protein and Amino Acid Function,* Bigwood, E. Y., Ed., Pergamon Press, New York, 1972, 10, 475.
11. **Shaternikov, V. A., Vysotski, V. G., Yatsyshina, T. A., Baturin, A. K., and Safronova, A. M.,** Ways of increasing the biological value of vegetable proteins, *Vopr. Pitan.,* 6, 20, 1982.
12. **Pellett, P. L. and Young, V. R.,** Nutritional evaluation of protein foods, in *Food and Nutrition Bulletin,* United Nations University, Suppl. 4, 1980.
13. **Hopkins, D. T.,** Effects of variation in protein digestibility, in *Protein Quality in Humans: Assessment and In Vitro Estimation,* AVI Publishing, Westport, CT, 1981, 169.

Chapter 8

ISOLATED SOY PROTEIN IN INFANT FEEDING

Samuel J. Fomon and Ekhard E. Ziegler

TABLE OF CONTENTS

I. INTRODUCTION

Formulas providing protein from soy flour were introduced into the U.S. market in 1950 as an aid in management of infants allergic to or intolerant of milk. The formulas were pale tan in color and had a nutty odor. Parents complained that the formulas produced loose, somewhat malodorous stools that stained the diaper area. Formulas prepared with isolated soy protein (ISP) became commercially available in the 1960s and within 10 years almost completely replaced soy flour-based ISP formulas in the U.S. Formulas are only slightly darker in color than milk-based formulas and are nearly odorless.

Data on the percentage of formula-fed infants in the U.S. who are fed ISP formulas are not readily available. However, nonmilk-based formulas, a category that includes, besides ISP formulas, commercially prepared goat milk and formulas with nitrogen from casein hydrolysates, accounted for 27% of infant formula sales in 1987.[1] Because some of these so-called "hypoallergenic" formulas are consumed by children and adults rather than by infants, sales figures do not directly reflect the percentage of infants fed these formulas. Market research data[2] on infants fed "hypoallergenic" formulas as a percentage of formula-fed infants are presented in Figure 1 for ages 2 to 3 months and 5 to 7 months. Most of these "hypoallergenic" formulas are ISP formulas. We estimate that at present about 20% of formula-fed infants in the U.S. are fed ISP formulas.

Much of the current usage of ISP formulas appears to rest on the belief of some physicians that fussiness and regurgitation is less common with ISP formulas than with milk-based formulas. There is little basis for this view. Some usage is attributable to the absence of lactose from ISP formulas and rests on the belief that a lactose-free feeding is advantageous during and following recovery from diarrhea.

The Committee on Nutrition of the American Academy of Pediatrics[3] has presented recommendations for use of ISP formulas in infant feeding. The Committee considers such formulas suitable for term, but not for preterm infants.

A. COMPOSITION OF ISP FORMULAS

ISP formulas marketed in the U.S. consist of methionine-fortified isolated soy protein (2.7 to 3.0 g/100 kcal), a mixture of vegetable oils, a cornstarch hydrolysate and/or sucrose, and vitamins, minerals, taurine, carnitine, and one or more stabilizers (mono- and diglycerides, soy lecithin, carageenan). The isolated soy proteins are derived from defatted soybean flakes, using a slightly alkaline aqueous solution to extract the protein. The protein is precipitated by adjustment of the pH to the isoelectric point of 4.5. The isolated soy proteins contain at least 90% protein on a dry basis.

The formulas comply with the requirements for nutrient concentrations specified by the U.S. Food and Drug Administration (FDA, 1985),[4] as presented in Table 1. Most ISP formulas are sold as concentrated liquid products (133 kcal/dl) that merely require the addition of an equal volume of water before they are fed. A small percentage of sales, mostly to hospitals, is in the ready-to-use (67 kcal/dl) form. Formulas marketed in powder form are commonly used as supplemental feedings for breast-fed infants.

Until about 1980, a number of the minerals present in some ISP formulas were poorly suspended and tended to settle out in feeding bottles or in syringes and tubing used for intragastric infusions.[5] These formulas may have resulted in inadequate mineral intake. Inadequate delivery of minerals might, of course, adversely affect bone mineralization. In mineral balance studies, if quantitative delivery of the minerals to the infant is not assured, retention will be overestimated. Currently, marketed ISP formulas appear to have better mineral stability.[6] Because 20 to 30% of the phosphorus provided by ISP formulas is contained in the phytate and is largely unavailable to the infant, somewhat greater concentrations of phosphorus are needed in ISP formulas than in milk-based formulas.

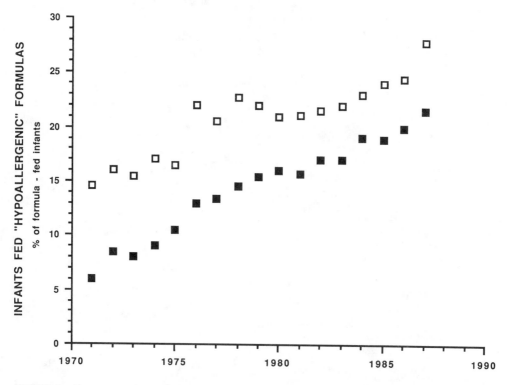

FIGURE 1. Percentage of formula-fed infants fed "hypoallergenic" formulas (see text) from 1971 through 1987. Black squares refer to infants between 2 and 3 months of age; open squares refer to infants between 5 and 7 months of age.[2]

B. ALUMINUM IN ISP FORMULAS

Concentrations of aluminum are known to be greater in ISP formulas than in milk-based formulas.[7] The aluminum in infant formulas is derived primarily from aluminum-contaminated mineral salts used in formula production. The extent to which aluminum is contributed by other components of formulas is under investigation.

A considerable body of evidence suggests that certain disorders of the central nervous system and skeleton commonly observed in patients with chronic renal disease are manifestations of aluminum toxicity.[8] The primary sources of the aluminum appear to be dialysis fluids, intravenously administered fluids, and aluminum-containing antacids administered as phosphate binders. In the face of poor renal function, it seems likely that the body burden of aluminum increases and is responsible for the clinical manifestations.

Elevated concentrations of aluminum were reported[9] in the brains of two infants with renal failure: a term infant who died at age 3 months and a preterm infant who died at age 1 month. Neither had experienced dialysis, had received fluids intravenously, or had been treated with phosphate binders or other aluminum-containing medications. The authors speculated that the source of aluminum responsible for elevated aluminum concentration in the brain was infant formulas. They reported that concentrations of aluminum in various formulas commercially available in the U.S. ranged from 124 to 316 ng/ml. The concentration in human milk was 4 ng/ml.

Although the health significance of aluminum contributed by infant formulas is not well established for infants with chronic renal disease, it seems prudent to use formulas as low as possible in aluminum content for such infants. For infants with normal renal function, there is currently relatively little basis for concern. However, additional study of this topic is needed before a confident statement can be made about the level of risk. It seems unlikely

TABLE 1
Nutrient Specifications for Infant
Formulas (Units per 100 kcal)

	Minimum	Maximum
Protein (g)	1.8	4.5
Fat (g)	3.3	6.0
Linoleic Acid (g)	0.3	—
Vitamin A (IU)	250	750
Vitamin D (IU)	40	100
Vitamin E (IU)	0.7	—
Vitamin K (µg)	4	—
Thiamine (µg)	40	—
Riboflavin (µg)	60	—
Vitamin B_6 (µg)	35	—
Vitamin B_{12} (µg)	0.15	—
Niacin[a] (µg)	250	—
Folic acid (µg)	4	—
Pantothenic acid (µg)	300	—
Biotin[b] (µg)	1.5	—
Vitamin C (mg)	8	—
Choline[b] (mg)	7	—
Inositol[b] (mg)	4	—
Calcium (mg)	60	—
Phosphorus (mg)	30	—
Magnesium (mg)	6	—
Iron (mg)	0.15	3.0
Zinc (mg)	0.5	—
Manganese (µg)	5	—
Copper (µg)	60	—
Iodine (µg)	5	75
Sodium (mg)	20	60
Potassium (mg)	80	200
Chloride (mg)	55	150

[a] Includes nicotinic acid and niacinamide.
[b] Required only for nonmilk-based infant formulas.

From Food and Drug Administration, 1985. Federal
Register 48187, Nov. 22, 1985.

that the diminished skeletal mineralization observed in term and especially in preterm infants
is related to the aluminum content of ISP formulas.[10]

II. NUTRITIONAL ADEQUACY FOR TERM INFANTS

The nutritional adequacy of ISP formulas has been explored in normal term infants[11-23]
and in term infants recovering from malnutrition[24] or diarrhea.[25]

A. GROWTH STUDIES
Growth was judged normal in all but two of the reports on ISP formulas. Cherry et al.[11]
reported less satisfactory growth by term infants fed an ISP formula than by control infants
fed a milk-based formula. However, in the report of this study, zinc is not listed in the table
of nutrients composition and one must assume that infants fed the formula received a zinc-
deficient diet. Kohler et al.[21] concluded that weight gain during the early months of life was
less rapid in infants fed an ISP formula than in breast-fed infants or infants fed a milk-based

formula. Unfortunately, only 13 infants were included in the group fed ISP formulas, and only size data (not gain data) are presented. Wiseman[14] reported 2-month weight gains of 20 males and 18 females less than 3 months of age, 23 males and 25 females 3 to 6 months of age, and 46 males and 38 females 6 to 12 months of age. Mean gains in weight and length were satisfactory as judged by reference data that we shall present later in this chapter.

In 1979, we[18] reported energy intakes and gains in weight and length from 8 to 112 d of age for 333 infants fed milk-based formulas (174 males and 159 females) and for 141 infants fed ISP formulas (74 males and 67 females). Gains in weight and length were similar with the two types of formulas, but energy intake (kcal kg^{-1} day^{-1}) was slightly greater by infants fed ISP formulas than by those fed milk-based formulas. The difference was not statistically significant. However, weight gain per unit of energy intake was less by infants fed ISP formulas than by those fed milk-based formulas: 5.52 vs. 5.74 g/100 kcal for males ($p < 0.01$), and 5.19 vs. 5.44 g/100 kcal for females ($p < 0.01$).

Our studies summarized in 1979 concerned formulas that contained more sucrose than those currently marketed. Feeding practices also differed. We have therefore summarized data from observations made after January 1, 1979. None of the infants was fed beikost (foods other than milk or formula) during the observation period from 8 to 112 days of age. The milk-based and ISP formulas were commercially available products or rather minor variations of commercially available products. As may be seen from Table 2, gains in weight (grams per day) and length (millimeters per day) were similar with the two types of formulas and were similar to values from our earlier report. Energy intake (kcal $\cdot$ kg^{-1} $\cdot$ day^{-1}) by infants fed ISP formulas was slightly lower than in our previous report and was almost identical to energy intake by infants fed milk-based formulas.

B. NITROGEN BALANCE

When infants receive similar intakes of nitrogen from two formulas that differ in the quality of the protein provided, retention of nitrogen may be greater from the formula with better protein quality.[20,24,26] We have previously reviewed nitrogen balance studies of infants fed ISP formulas,[18] and we have subsequently published additional data.[20] In summary, studies with methionine-fortified ISP fed at levels similar to those of commercially available ISP formulas provide no suggestion of inadequacy of the protein. However, as an index of protein quality, we consider nitrogen balance to be less sensitive than growth rate or serum concentrations of albumin and urea nitrogen.[20]

C. SERUM ALBUMIN AND UREA NITROGEN

In normal infants fed adequate diets, serum concentrations of albumin demonstrate predictable increases during the early months of life.[20] When the protein quantity or quality of the diet is inadequate, serum concentrations of albumin may be low. For example, when we fed experimental formulas providing ISP unfortified with methionine at levels of 2.2 or 2.6 g/100 kcal, serum concentrations of albumin were less when these same formulas were fed with a supplement of methionine.[20] Commercially available ISP formulas in the U.S. provide generous concentrations of methionine-fortified ISP. When these formulas have been fed to term infants, serum albumin concentrations have been judged to be normal.[12,13,17,18,20] With human infants,[20] as with various animals,[27] protein quality of the diet may be reflected in the serum concentration of urea nitrogen. With similar intakes of nitrogen, the protein of poorer quality may be associated with a greater concentration of urea nitrogen. We were able to demonstrate an effect of methionine supplementation of experimental ISP formulas on serum concentration of urea nitrogen at protein concentrations of 2.2 g/100 kcal, but not at greater protein concentrations.[20]

D. MINERAL BALANCE

As already mentioned, mineral balance studies are prone to yield falsely high retentions

TABLE 2
Energy Intake and Growth of Term Infants Fed Milk-Based or ISP Formulas

	Males				Females			
	Milk-based formulas (57)[a]		ISP formulas (46)		Milk-based formulas (46)[a]		ISP formulas (55)	
	Mean	SD	Mean	SD	Mean	SD	Mean	SD
Age 8 to 41 d								
Energy intake								
kcal kg^{-1} d^{-1}	119	13.5	117	13.7	116	12.4	118	13.0
Gain in weight								
g/d	40.6	6.8	38.6	9.4	34.3	7.2	33.9	7.2
g/100 kcal	8.04	0.91	7.75	1.51	7.33	0.95	7.20	1.32
Gain in length								
mm/d	1.35	0.28	1.28	0.20	1.25	0.15	1.25	0.20
Age 42 to 111 d								
Energy intake								
kcal kg^{-1} d^{-1}	100	7.1	100	8.5	101	7.9	101	9.4
Gain in weight								
g/d	27.9	5.7	28.0	6.2	25.0	6.2	25.6	5.2
g/100 kcal	4.61	0.78	4.71	0.77	4.47	0.84	4.62	0.72
Gain in length								
mm/d	0.99	0.11	1.01	0.11	0.90	0.11	0.98	0.11
Age 8 to 111 d								
Energy intake								
kcal kg^{-1} d^{-1}	106	7.4	106	7.7	106	8.1	107	9.4
Gain in weight								
g/d	32.1	5.2	31.4	5.8	28.0	5.6	28.3	4.4
g/100 kcal	5.61	0.73	5.61	0.79	5.30	0.70	5.38	0.68
Gain in length								
mm/d	1.11	0.13	1.10	0.10	1.01	0.09	1.06	0.09

[a]   Number of subjects.

when mineral intakes are overestimated because a portion of the minerals settles out and is not delivered to the infant. With similar concentrations of calcium and phosphorus in an experimental milk-based formula and an experimental ISP formula, we demonstrated greater retention of these minerals from the ISP formula.[28] It is unlikely that intake of minerals was overestimated in that study. However, some settling out of minerals may have occurred during the prebalance periods at home, and the higher mineral retentions during balance studies might be explained on that basis.

E. BONE MINERAL CONTENT
Steichen and Tsang[23] studied normal-term infants fed an ISP formula or a milk-based formula from birth to 1 year of age. Bone mineral content, measured by single-beam photon absorptiometry of the distal radius, was greater in infants fed the milk-based formula at ages 3, 6, 9, and 12 months. This study was undertaken before changes were made in the mineral suspension of the ISP formula.

Chan et al.[22] studied normal-term infants fed ISP formulas during the first 4 months of life and included breast-fed infants as controls. At ages 2 months and 4 months, bone mineral content was greater in the breast-fed infants. There were no significant differences in serum concentrations of calcium, magnesium, or phosphorus nor in activity of alkaline phosphatase.

Hillman et al.[29] followed three groups of normal term infants from birth to 1 year of age. Bone mineral content did not differ at any age among breast-fed infants, infants fed a milk-based formula, and infants fed an ISP formula. Although serum concentration of phosphorus at 6 months of age was significantly lower in breast-fed infants than infants fed either of the formulas, activity of alkaline phosphatase was similar in the three groups. The discrepancy between the results of Hillman et al.[29] and those of Chan et al.[22] is unexplained. The same ISP formula with calcium and phosphorus, presumably in stable suspension, i.e., without the problem of settling out of minerals, was used in both studies.

Using a radiographic technique, Kohler et al.[21] estimated metacarpal cortical thickness and skeletal maturation at 3 and 6 months of age in breast-fed infants, in infants fed a milk-based formula, and in infants fed an ISP formula. At 3 months of age, infants fed the ISP formula had significantly thinner metacarpal cortical bone and significantly smaller carpal ossification centers than breast-fed infants and infants fed the milk-based formula. By 6 months of age these differences were diminished and no longer statistically significant.

III. NUTRITIONAL ADEQUACY FOR PRETERM INFANTS

The nutrient requirements of small preterm infants are quite different from those of term infants.[30] It is recognized that human milk must be supplemented with several nutrients to meet nutritional needs of small preterm infants,[31-34] and specially designed formulas are required. Currently available ISP formulas do not appear to meet the needs of small preterm infants.[3]

A. GROWTH AND SERUM CHEMICAL INDICES OF PRETERM INFANTS

Few satisfactory studies are available concerning the growth of preterm infants, especially small preterm infants, fed ISP formulas. Naude et al.[35] studied infants with birth weights between 1500 and 1800 g. Twenty infants were fed a whey-predominant cow milk formula and 20 were fed an ISP formula. Weight gain and serum albumin concentrations were significantly lower in infants fed the ISP formula than in those fed the whey-predominant milk based formula. Shenai et al.[36] studied infants with birth weights less than 1530 g. Ten infants were fed an ISP formula and nine were fed a milk-based formula. Gains in weight, length, and head circumference were similar and there were no significant differences in serum concentrations of albumin and total protein.

Hall et al.[37] studied preterm infants with birth weights less than 1500 g. They compared 17 infants fed an ISP formula with 15 infants fed a whey-predominant milk-based formula. Weight gain and serum albumin concentrations were significantly lower in infants fed the ISP formula than in those fed the milk-based formula.

B. NITROGEN AND MINERAL BALANCE

Shenai et al.[36] carried out 96-h metabolic balance studies with small preterm infants. Despite significantly higher nitrogen intake with the ISP formula, retention of nitrogen was significantly less than with the milk-based formula (62.2% of intake vs. 73.1% of intake). There were no significant differences in calcium balance, but percentage of absorption and urinary excretion of phosphorus were significantly lower in infants fed the ISP formula than in those fed the milk-based formula.

C. METABOLIC BONE DISEASE (RICKETS OF PREMATURITY)

Naude et al.[35] reported that serum concentration of phosphorus was significantly less and alkaline phosphatase activity was significantly greater in preterm infants fed an ISP formula than in those fed a whey-predominant cow milk formula. Somewhat similar findings were reported by Shenai et al.[36] Hall et al.[37] also observed significantly lower serum concentrations of calcium and phosphorus in infants fed an ISP formula than in those fed a whey-predominant formula. Alkaline phosphatase activity was not significantly greater. The serum chemical changes observed in these studies are characteristically observed in preterm infants with metabolic bone disease, but in none of the studies was an attempt made to evaluate bone status.

Kulkarni et al.[38] noted the frequent presence of rickets on radiographs from small preterm infants fed ISP formulas. Callenbach et al.[39] reported a retrospective review of 125 small preterm infants fed ISP formulas, supplemented in most cases with calcium and vitamin D. Radiographic evidence of rickets was present in 32% of infants. Serum concentrations of phosphorus were less and activity of alkaline phosphatase was greater in infants with radiographic evidence of rickets than in those without.

IV. SUMMARY

In the U.S., ISP formulas have become widely accepted, notwithstanding the fact that only a small percentage of normal infants are likely to benefit by being fed ISP formula rather than milk-based formulas. Currently, about 20% of formula-fed infants receive ISP formulas.

Although some earlier studies had raised questions about the ability of ISP formulas to support adequate growth of normal-term infants, the available data concerning growth and serum chemical indices overwhelmingly document the nutritional adequacy of ISP formulas for term infants. Whether or not bone mineral content of infants fed ISP formulas is less than that of breast-fed infants and infants fed milk-based formulas is not firmly established. A decrease in bone mineral content might result from inadequate delivery of minerals to the infant (settling out of calcium and phosphorus), lower availability of these minerals from ISP formulas, or to yet unidentified factors in ISP formulas.

ISP formulas are nutritionally inadequate for preterm infants. In these infants, growth performance, serum chemical values, and nitrogen balance data document the nutritional inadequacy of ISP, presumably because of the relatively low concentration of cystine and perhaps of methionine. Furthermore, a high percentage of preterm infants fed ISP formulas develop metabolic bone disease (rickets of prematurity).

REFERENCES

1. **Nielsen, A. C.,** (Personal communication), A. C. Nielsen Co., Northbrook, IL, 1988.
2. **Martinez, G. A.,** (Personal communication), Ross Laboratories, Columbus, OH, 1988.
3. Committee on Nutrition, American Academy of Pediatrics, Soy protein formulas: recommendations for use in infant feeding, *Pediatrics,* 72, 359, 1983.
4. Food and Drug Administration Rules and Regulations: nutrient requirements for infant formulas, *Fed. Regist.,* 21 CFR Part 107, 50, 45106, 1985.
5. **Bhatia, J. and Fomon, S. J.,** Formulas for premature infants: fate of the calcium and phosphorus, *Pediatrics,* 72, 37, 1983.
6. **Bhatia, J.,** Formula fixed, *Pediatrics,* 75, 800, 1985.
7. **Sedman, A. B., Klein, G. L., Merrit, R. J., Miller, N. L., Weber, K. O., Gill, W. L., Anand, H., and Alfrey, A. C.,** Evidence of aluminum loading in infants receiving intravenous therapy, *N. Engl. J. Med.,* 312, 1337, 1985.
8. **Polinsky, M. S. and Gruskin, A. B.,** Aluminum toxicity in children with chronic renal failure, *J. Pediatr.,* 105, 758, 1984.
9. **Freundlich, M., Abitbol, C., Zilleruelo, G., and Strauss, J.,** Infant formula as a cause of aluminum toxicity in neonatal uremia, *Lancet,* 527, 1985.
10. **Koo, W. W. K. and Kaplan, L. A.,** Aluminum and bone disorders: with specific reference to aluminum contamination of infant nutrients, *J. Am. Coll. Nutr.,* 7, 199, 1988.

11. **Cherry, F. F., Cooper, M. D., Stewart, R. A., and Platou, R. V.,** Cow versus soy formulas: comparative evaluation in normal infants, *Am. J. Dis. Child.,* 115, 677, 1968.

12. **Bates, R. D., Barrett, W. W., Anderson, D. W., and Saperstein, S.,** Milk and soy formulas: a comparative study, *Ann. Allergy,* 26, 577, 1968.

13. **Cowan, C. C., Brownlee, R. C., DeLoache, W. R., Jackson, H. P., and Matthews, J. P.,** A soy protein isolate formula in the management of allergy in infants and children, *South. Med. J.,* 62, 389, 1969.

14. **Wiseman, H. J.,** Comparison of two soy protein isolate infant formulas, *Ann. Allergy,* 29, 209, 1971.

15. **Dean, M. E.,** A study of normal infants fed a soya protein isolate formula, *Med. J. Aust.,* 1, 1289, 1973.

16. **Fomon, S. J., Thomas, L. N., Filer, L. J., Anderson, T. A., and Bergmann, K. E.,** Requirements for protein and essential amino acids in early infancy, *Acta Paediatr. Scand.,* 62, 33, 1973.

17. **Jung, A. L. and Carr, S. L.,** A soy protein formula and a milk-based formula: a comparative evaluation in milk-tolerant infants showed no significant nutritional differences, *Clin. Pediatr.,* 16, 982, 1977.

18. **Fomon, S. J. and Ziegler, E. E.,** Soy protein isolates in infant feeding, in *Soy Protein and Human Nutrition,* Wilcke, H. L., Hopkins, D. T., and Waggle, D. H., Eds., Academic Press, New York, 1979.

19. **Fomon, S. J., Ziegler, E. E., Filer, L. J., Nelson, S. E., and Edwards, B. B.,** Methionine fortification of a soy protein formula fed in infants, *Am. J. Clin. Nutr.,* 32, 2460, 1979.

20. **Fomon, S. J., Ziegler, E. E., Nelson, S. E., and Edwards, B. B.,** Requirement for sulfur-containing amino acids in infancy, *J. Nutr.,* 116, 1405, 1986.

21. **Köhler, L., Meeuwisse, G., and Mortensson, W.,** Food intake and growth of infants between six and twenty-six weeks of age on breast milk, cow's milk formula, or soy formula, *Acta Paediatr. Scand.,* 73, 40, 1984.

22. **Chan, G. M., Leeper, L., and Book, L. S.,** Effects of soy formulas on mineral metabolism in term infants, *Am. J. Dis. Child.,* 141, 527, 1987.

23. **Steichen, J. J. and Tsang, R. C.,** Bone mineralization and growth in term infants fed soy-based or cow milk-based formula, *J. Pediatr.,* 110, 687, 1987.

24. **Graham, G. G., Placko, R. P., Morales, E., Acevedo, G., and Cordano, A.,** Dietary protein quality in infants and children: isolated soy protein milk, *Am. J. Dis. Child.,* 120, 419, 1970.

25. **Leake, R. D., Schroeder, K. C., Benton, D. A., and Oh, W.,** Soy-based formula in the treatment of infantile, *Am. J. Dis. Child.,* 127, 374, 1974.

26. **Kaye, R., Barness, L. A., Valyasevi, A., and Knapp, J.,** Nitrogen Balance Studies of Plant Proteins in Infants, in The Committee on Protein Malnutrition, Food and Nutrition Board and The Nutrition Study Section, National Institutes of Health: Progress in Meeting Protein Needs of Infants and Preschool Children, Publication 843, National Academy of Sciences, National Research Council, Washington, D.C., 1961.

27. **Bodwell, C. E.,** Biochemical indices in humans, in *Evaluation of Proteins for Humans,* Bodwell, C. E., Ed., AVI Publishing, Westport, CT, 1977, 119.

28. **Ziegler, E. E., Fomon, S. J., Edwards, B. B., and Nelson, S. E.,** Mineral absorption from soy-based and milk-based infant formulas, *Am. J. Clin. Nutr.,* 35, 823, 1982 (abstr.).

29. **Hillman, L. S., Chow, W., Salmons, S. S., Weaver, E., Erickson, M., and Hansen, J.,** Vitamin D metabolism, mineral homeostatis, and bone mineralization in term infants fed human milk-based formula or soy-based formula, *J. Pediatr.,* 112, 864, 1988.

30. **Ziegler, E. E., Biga, R. L., and Fomon, S. J.,** Nutritional requirements of the premature infant, in *Textbook of Pediatric Nutrition,* Suskind, R. M., Ed., Raven Press, New York, 1981.

31. **Ziegler, E. E.,** Nutritional management of the premature infant, *Perinatol.-Neonatol.,* 9, 11, 1985.

32. **Schanler, R. J., Garza, C., and Nichols, B. L.,** Fortified mothers' milk for very-low-birth-weight infants: results of growth and nutrient balance studies, *J. Pediatr.,* 107, 437, 1985.

33. **Rönnholm, K. A. R., Perheentupa, J., and Siimes, M. A.,** Supplementation with human milk protein improves growth of small premature infants fed human milk, *Pediatrics,* 77, 649, 1986.

34. **Carey, D. E., Rowe, J. C., Goetz, C. A., Horak, E., Clark, R. M., and Goldberg, B.,** Growth and phosphorus metabolism in premature infants fed human milk, fortified human milk, or special premature formula, *Am. J. Dis. Child.,* 141, 511, 1987.

35. **Naudé, S. P. E., Prinsloo, J. G., and Haupt, C. E.,** Comparison between a humanized cow's milk and a soy product for premature infants, *S. Afr. Med. J.,* 55, 982, 1979.

36. **Shenai, J. P., Jhaveri, B. M., Reynolds, J. W., Huston, R. K., and Babson, S. G.,** Nutritional balance studies in very-low-birth-weight infants: role of soy formula, *Pediatrics,* 67, 631, 1981.

37. **Hall, R. T., Callenbach, J. C., Sheehan, M. B., Hall, F. K., Thilbeault, D. W., Kurth, C. G., and Bowen, S. K.,** Comparison of calcium- and phosphorus-supplemented soy isolate formula with whey-predominant premature formula in very-low-birth-weight infants, *J. Pediatr. Gastroenterol. Nutr.,* 3, 571, 1984.

38. **Kulkarni, P. B., Hall, R. T., Rhodes, P. G., Sheehan, M. B., Callenbach, J. C., Germann, D. R., and Abramson, S. J.,** Rickets in very-low-birth-weight infants, *J. Pediatr.,* 97, 249, 1980.

39. **Callenbach, J. C., Sheehan, M. B., Abramson, S. J., and Hall, R. T.,** Etiologic factors in rickets of very-low-birth-weight infants, *J. Pediatr.,* 98, 800, 1981.

Chapter 9

THE USE OF AN ISOLATED SOY PROTEIN FORMULA FOR NOURISHING INFANTS WITH FOOD ALLERGIES

Kaleria S. Ladodo and Tatyana E. Borovick

Over the last decade, a significant increase in allergic disorders has been observed throughout the world, particularly among children.[1-4] The vast majority of these allergic pathologies are food related and range in frequency according to data from various researchers from 15 to 60%.[5-7] The greatest incidence of food allergies is found among very young children and is mainly linked to a sensitivity towards proteins in cow milk.

The primary clinical manifestations of food allergy in children under 1 year of age are skin disorders in the form of cheek rash, urticaria, and true eczema. Disorders also occur in the gastrointestinal tract, such as regurgitation, vomiting, and diarrhea. Often, a combination of skin and intestinal allergy symptoms are observed. Sensitivity to foods in older children may manifest itself in the respiratory organs in the form of allergic rhinosinusitis, obstructive bronchitis, and bronchial asthma. In severe cases this can result in retarded growth and anaphylaxis.

For detecting the susceptibility to food allergies, various diagnostic methods are used, among which the most sensitive and informative are the radioimmunosorbent test (RIST) for detecting total IgE antibodies and the radioallergosorbent test (RAST) for detecting IgE-specific antibodies.[8-10]

Another valuable diagnostic tool is the erythrocyte hemagglutination (RPGA) antibody test for casein, serum alpha-lactoalbumin and beta-lactoglobulin, and bovine serum albumin. Such a set of food antigens permits the detection not only of the sensitizing action of cow milk, but also the diagnosis of cause-specific fractions of the protein. In addition, this method does not require special equipment, which makes it widely applicable in pediatric practice.[11,12]

Diet therapy, based on the elimination method which excludes cause-specific allergens and products with high allergy-causing potential, is the primary treatment of food allergies. This method of treatment presents serious difficulties for children under 1 year of age in cases where they are sensitive to cow milk antigens, which is the main type of food for these children. For this reason, many researchers are seeking a way to provide an adequate diet for such children which will exclude cow milk from the diet.

At the Nutrition Institute of the Academy of Medical Sciences (AMS) of the U.S.S.R. and the Institute of Pediatrics AMS, intensive work has been conducted on diet therapy with food allergies in very young children. We observed 185 children ranging in age from 3 weeks to 18 months. Of these, 142 of them were diagnosed as having atopic dermatitis of an acute or semiacute nature, and 43 were seen to have a complex of allergic skin reactions with gastrointestinal problems in the form of serious diarrhea. The majority (62%) had illnesses of moderate seriousness, 29% were in serious condition, and 9% were only slightly affected. The seriousness of their condition was determined by the degree of manifestation of infiltrative and infective changes on the skin, their degree of spreading, the intensity of itching, and the absence of effects of the therapy conducted. Among accompanying disorders were observed first- and second-degree rickets (51%) and first- and second-degree hypotrophy (33%).

The PRIST and RAST tests were used to diagnose food allergies. In 60.7% of the patients we discovered an elevation of the average level of class E immunoglobulins (292 $\pm$ 15.9 U/l) in comparison with the norm for this age group (16.5 $\pm$ 9.8 U/l). Positive tests ($++$) for cow milk protein were observed in 56% of the patients, and highly positive

(+ + +) results were seen in 25% of the children, which indicated their high sensitivity to milk antigens.

We conducted a comparative study of the effectiveness of a soy formula, Isomil® (U.S.), and traditionally used domestic sour-milk products in diet therapy for children with allergies to cow milk. The basis for using sour-milk formulas is data which indicate the potential of various strains of milk-fermenting bacteria to partially break down casein with the formation of free amino acids as well as hydrolysis of lactose. In addition, fermented milk is reported to stimulate the secretion of digestive juices, decrease the fermentation process in the intestine, and suppress and eliminate disease-producing microorganisms, thus permitting improvement in the digestive processes and a decrease in skin rashes. We used acidified formula "Malyutka", which employs a strain of Lactobacillus acidified for fermentation, that is physiologically natural for children and has a high proteolytic and antibacterial effect.[13,14]

The "milk-free" formula Isomil®, whose chemical composition is close to mother's milk, is prepared using isolate soy protein of Purina® Proteins 710 (Protein Technologies International). The use of soy protein as the basis for a diet formula is due to its high content of essential amino acids in comparison with other protein sources. In its amino acid makeup, soy is close to products of animal origin. The formula Isomil®, in addition to isolate soy protein, contains corn, cocoa, soy oil, corn syrup, sucrose, tapioca starch, vitamins, and mineral salts. With a view to increasing the biological value of the isolated soy protein in the product, L-methionine is added. The absence of milk protein and lactose in Isomil® makes it suitable for patients suffering from food allergies and secondary lactase deficiency.

The main criteria for evaluating the effectiveness of diet therapy using acidified-formula Malyutka and Isomil® formula were: improvement of skin conditions, normalization of the stool and the nutritional status of the child, and a positive improvement for individual immunological, microbiological, and biochemical indices.

These formulas were included in the child's diet with great care, by initially adding 3 to 5 g, and then gradually increasing the daily volume in 20-g increments until the total amount reached 400 to 800 g (depending on the child's age) after 10 to 14 d. The inclusion of acidified formula (Malyutka) and Isomil® in the diet of the majority of the patients with food allergies was carried out without complications. A tendency toward improvement of the general condition of children with atopic dermatitis by the gradual disappearance of spots, a decrease in the rash, and an improvement in skin elasticity was observed in 89% of the children after 2 weeks from the time the products were introduced. Clinical remission of illness was recorded after 8 to 10 weeks on treatment.

With patients having a complex of skin and gastrointestinal problems, a decrease in intestinal problems was observed after 5 to 6 d of using these formulas, and normalization occurred by the 10th to 12th day. The disappearance of infiltrative and infectious changes of the skin was noted 6 to 8 weeks from the beginning of diet treatment using isolated soy protein formula or sour-milk products. Our studies also demonstrated that the use of these formulas also had a positive influence on the indices of physical development for the children (Figures 1 and 2). The average daily increase in body weight was 28.0 ± 3.5 g for those using Isomil® and 27.5 ± 2.9 g for those children on acidified-formula Malyutka, which is a normal rate of growth for infants.

As an example, we cite the case of 7-month-old Alyosha M., a child of a third pregnancy, which proceeded with toxicosis during the first and second semesters. The clinical diagnosis was atopic dermatitis of an acute form, with an allergy to cow milk proteins with first-degree hypotrophy. Birth weight was 3900 g, length 54 cm. Allergological history showed that the mother had eczema and a red rash. The child was breast fed the first 6 weeks. From the age of 2 months, after beginning feeding with the formula Krepysh, the first manifestations of exudative catarrhal diathesis were present. By 2.5 months the condition had turned into eczema and had spread over the entire body.

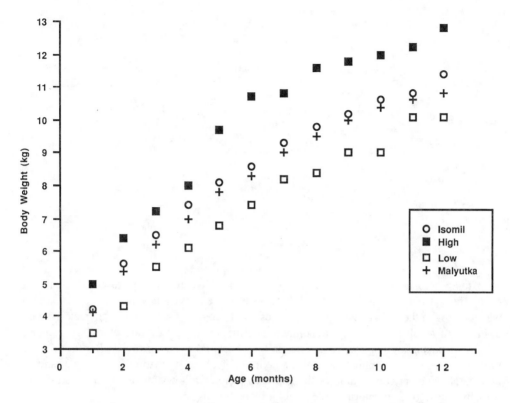

FIGURE 1.   Changes in body weight of infants fed Isomil or Malyutka.

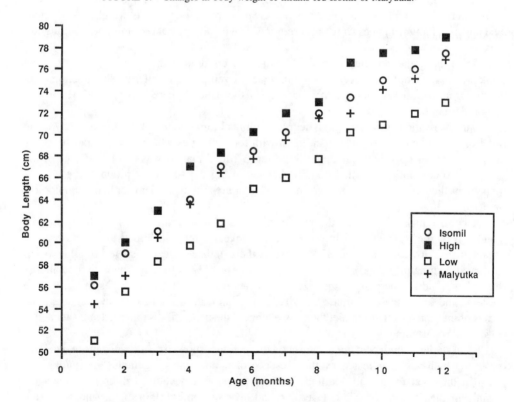

FIGURE 2.   Changes in body length of infants fed Isomil or Malyutka.

TABLE 1
Nonmilk Diet

Feeding time	Food	Amount
0600	Isomil® formula	200 ml
1000	Rice porridge with water	130 g
	Isomil® formula	50 ml
	Apple sauce	40 g
1400	Vegetable puree	150 g
	5 g Vegetable oil	
	Meat puree	30 g
	Apple juice	50 ml
1800	Vegetable puree	130 g
	5 g Vegetable oil	
	Meat puree	30 g
	Apple sauce	40 g
2200	Isomil® formula	200 ml

Atopic dermatitis in the form of numerous loci of hyperemia with rashes was present on the face, trunk, and extremities. A high level of sensitivity to cow milk proteins also was found with the RAST test (+ + +). Intestinal bacteriological changes characterized by disorders of aerobic and anaerobic components (the absence of bifidobacteria, an increase in plasmacoagulating staphylococci and proteus bacteria) were observed. Based on these symptoms and tests, the child was given a nonmilk diet using the Isomil® formula as shown in Table 1. Nutrient composition of the diet used were as follows: protein, 3.3 g; fat, 5.9 g; carbohydrates, 14.2 g; calories, 123 kcal/kg of body mass per day.

Feeding of the child using the Isomil® formula continued for 2.5 months. At the end of this period, the child's skin problems had cleared up completely, his weight was normal, and positive immune indices were observed. RAST with allergen of cow milk became slightly positive.

When using the acidified-formula Malyutka in children with acute sensitivity to cow milk proteins, we did not succeed in obtaining improvement in the allergic skin phenomenon or in normalizing body weight. In a number of the patients there was an exacerbation of atopic dermatitis.

Our observations showed a high degree of clinical improvement from diet therapy using the isolated soy protein product in children with food allergies which were characterized by a high degree of sensitivity to cow milk proteins. Blood plasma was practically free of the antibody to cow milk protein. In addition, when using the sour-milk product, these types of patients did not experience positive changes either in the skin problems or in immunological indices.

A comparative analysis of the results from tests of several indices of the functional condition of the pancreas and intestine did not reveal a relationship to the foods used in infant diets. However, a direct link between these indices and the stages of illness was observed. During the period of increased severe illness, 69% of the children with food allergy had edema and an increase in the size of the pancreas, significant changes in the trypsin system, and an increase in lipase activity, which indicate a breakdown in the functional status of the pancreas. During diet therapy, edema of the pancreas disappeared and a tendency toward normalized activity was observed.

Fecal studies, conducted upon admission to the clinic, showed disturbance in the digestive process which was manifested by the presence of neutral fats, vegetable, and undigested meat in the fecal matter. In addition, 85% of the infants developed microflora changes, mainly in the form of a series of aerobic and anaerobic microflora disturbances. During remission, a positive change in the fecal pattern was observed as well as indices which characterized microflora of the intestine in the form of an increased content of bifidobacteria and lactobacteria and a decrease in lactose-negative and hemolyzing strains among enterobacteria and proteus bacteria.

We have determined that in infants with atopic dermatitis, the excretion of carbohydrates in the stool is similar to normal infants. However, in patients with a complex of skin and gastrointestinal manifestations of food allergies, there was noted an increased excretion of carbohydrates in fecal matter (2.48%), as well as lactose (1.59%), galactose (0.31%), and xylose (0.58%). The frequency of occurrence of lactose in fecal matter was 98%, galactose 62.5%, and xylose 75%. The content and percentage for the presence of other sugars did not differ from the same indices in patients with atypical dermatitis and in healthy infants. Intestinal symptoms and increased excretion of lactose with fecal matter in this group revealed the presence of secondary lactase insufficiency. During the diet therapy, the excretion of carbohydrates in feces gradually returned to the age-normal levels.

Thus, in infants under 1 year of age who had food allergies which developed as a result of sensitivity to milk proteins, the most effective diets were composed of acidified-formula Malyutka and the Isomil® product. The Isomil® formula, based on an isolated soy protein, is balanced in all its nutrients and had a favorable influence on the clinical progress of the illness and the physical status of patients suffering from atopic dermatitis, which was characterized by acute sensitivity to cow milk protein. This permits the conclusion that its use, along with special forms of sour-milk formulas made from products prepared with isolates of soy protein, is justified for diet therapy with infants having food allergies, especially those with acute sensitivity to milk antigens.

REFERENCES

1. **Vorontsov, I. M. and Matalygina, O. A.**, Illnesses linked to food sensitivity, *Leningrad*, 271, 1986.
2. **Studenikin, M., Ya. and Sokolova, T. S.**, *Allergic Disorders in Children*, Meditsina, Moscow, 1986, 287.
3. **Bahna, S. L. and Heiner, D. C.**, *Allergies to Milk*, XII, Grune & Stratton, New York, 1980, 202.
4. **Crawford, L. A.**, Allergy diets in *Allergic Diseases of Infancy, Childhood and Adolescence*, W. B. Saunders, Philadelphia, 1980, 394.
5. **Revyakina, V. A.**, Clinical and Immunological Symptoms of Food Allergies in Children, Dissertation, Moscow, 1983.
6. **Eyubova, A. A.**, Clinical-Pathogen Variants of Food Allergy in Children and Its Prevalence in the Azerbaijan SSR, Dissertation, Baku, 1981.
7. **McLaanhlan, P. and Coombs, R.**, Latent anaphylactic sensitivity of infants to cow's milk proteins, *Clin. Allergy*, 13, 1, 1983.
8. **Eletskiy, A., Yu.**, Immunological disorders in eczema in children and parents, *Pediatriya*, 12, 36, 1981.
9. **Partsialis, Ye., M.**, Clinical Significance of Studies of Digestive Secretions in Children with Skin and Respiratory Allergy Symptom, Dissertation, Moscow, 1983.
10. **Jung, A. L. and Carr, S. L.**, A soy protein formula and milk-based formula, *Clin. Pediatr.*, 16, N.II, 982, 1977.
11. **Kuvayeva, I. B., Zakharova, N. V., Orlova, N. G., and Veselova, O. L.**, The functional condition of the immunological system and the digestive tract children with food allergies, *Vopr. Pitan.*, 32, 33, 1980.
12. **Iuvayeva, I. B., Basova, N. N., and Veselova, O. L.**, Discovery of antibodies to food antigens in a passive hemagglutination reaction with a series of entrocyte diagnostics of varying specificity, *Immunologiya*, 4, 85, 1982.
13. **Borovik, T. E.**, Diet Therapy in Allergo-Dermatoses in Children Under One Year of Age, Dissertation for medical sciences doctorate, 1983.
14. **Ivanova, L. N., Barashneva, S. M., Mamonova, L. G., and Oleneva, I. V., et al.**, The Creation and Clinical Approbation of Liquid Acidified Formulas in Healthy and Sick Infants Under One Year of Age and Their Improvement, The book treats the theoretical and practical aspects of human nutrition studies, Ph.D. thesis, Vol. 3, 207, 1980.

Chapter 10

SOY PROTEINS AS AMINO ACID AND PROTEIN SOURCES FOR PRESCHOOL-AGE CHILDREN

Benjamin Torun

TABLE OF CONTENTS

I. INTRODUCTION

New sources of protein foods are needed for optimal nutrition and health of populations with diets that are lacking or marginal in terms of total nitrogen and essential amino acid contents. This need grows with the increase in cost of traditional protein sources and with the changes in eating habits induced by urbanization, migrations, and cultural changes. Vegetables and soy, in particular, are economic sources of protein that can be used to increase the supply of dietary protein and/or improve the balance of essential amino acids through combinations that complement the limiting amino acids of the various protein sources in the diet.[1,2]

After the infant, the child of preschool age is the most sensitive in terms of essential amino acid requirements. The primary needs are for tissue maintenance and body functions. Growth requires a relatively small, but highly important proportion of the total amino acid needs. Where food resources are limited or its distribution is unequal, protein and amino acid restrictions during this critical age can result in a reduction of body size, often accompanied by functional limitations.

II. ESSENTIAL AMINO ACID REQUIREMENTS AND RECOMMENDATIONS

Prior to 1981, the Food and Agriculture Organization/World Health Organization (FAO/WHO) Expert Committees on protein requirements did not have available adequate experimental data of the essential amino acid requirements of preschool-age children. Beyond early infancy, the only existing information was from the studies of Nakagawa and co-workers[3,4] with children 10 to 12 years old. The 1971 committee[5] estimated the requirements of preschool children by interpolation between those of infants and the 10- to 12-year olds. It also recognized that the protein value of a food could be estimated from its essential amino acid content in relation to an amino acid pattern of reference, which was based on the estimated requirements of preschool children. In 1975, it was further recommended that this protein value should be corrected by the digestibility of the protein source in relation to that of milk and egg as reference proteins.[6]

The lack of experimental evidence on the amino acid requirements of preschool children led to a series of studies conducted by Torun et al.[7] and Pineda and co-workers[8] at the Institute of Nutrition of Central America and Panama (INCAP). They used semielemental diets based on a core level of cow milk and synthetic amino acids. The amounts of each specific essential amino acid under study were gradually increased or decreased. The children's requirements were established from serial analyses of free amino acids in plasma, nitrogen balance, and urinary urea/creatinine ratio. Further studies using cow milk and isolated soy proteins to determine protein requirements of similar children[9,10] confirmed the findings and suggested that the amounts of essential amino acids derived from the studies with semielemental diets could in fact be considered as "recommended" or "safe levels" of intake.[7] Those amounts are shown in Table 1.

III. SUGGESTED PATTERN OF AMINO ACID REQUIREMENTS

The 1981 FAO/WHO/United Nations University (UNU) Expert Committee on energy and protein requirements[11] accepted the recommendations proposed by Torun and co-workers.[7] The daily safe level of protein intake for children 2 to 5 years old was taken as 1.10 g reference (milk) protein per kilogram. Dividing the recommended levels of essential amino acids by 1.10 gave origin to the amino acid pattern of requirements suggested for preschool children (Table 2). The "provisional pattern" suggested in 1973 and the amino acid composition of soy protein are shown for comparison.

TABLE 1
Recommended Essential Amino Acid
Intakes for Preschool Children

Amino Acid	Recommended intake (mg/kg/d)
Histidine[a]	
Isoleucine	31
Leucine	73
Lysine	64
Methionine + cystine	27
Phenylalanine + tyrosine	69
Threonine	37
Tryptophan	12.5
Valine	38

Note: Must be corrected for digestibility of protein source relative to milk.

[a] Requirement for histidine not measured.

From Torun, B., Pineda, O., Viteri, F. E., and Arroyave, G., in *Protein Quality in Humans: Assessment and In Vitro Estimation,* Bodwell, C. E., Adkins, J. S., and Hopkins, D. T., Eds., AVI Publishing, Westport, CT, 1981, 374. With permission.

TABLE 2
Current and Previously Suggested Patterns of Essential Amino Acid
Composition of Proteins (Milligrams Amino Acid Per Gram Protein)

Amino acid	Current pattern (1985)[a]	Provisional pattern (1973)	Isolated soy protein[b]
Histidine	(19)[c]	—	26
Isoleucine	28	40	49
Leucine	66	70	82
Lysine	58	55	63
Methionine + cystine	25	35	26
Phenylalanine + tyrosine	63	60	90
Threonine	34	40	38
Tryptophan	11	10	14
Valine	35	50	50

[a] From FAO/WHO/UNU,[11] based on the recommended intake shown in Table 1 and a safe level of protein intake of 1.10 g reference protein per kilogram per day.

[b] Source: Protein Technologies International, St. Louis, MO.

[c] Histidine requirement not measured in preschool children. Value in parentheses was interpolated from data in infants and estimates in adults.

This pattern, expressed in terms of milligrams of amino acid per gram of protein, has become the standard of comparison to assess the protein quality of a food by calculating its amino acid score (according to the most limiting amino acid, i.e., the one with the largest deficit) and correcting it by the digestibility of the protein.[11] A protein with an amino acid composition that coincides with or surpasses the amino acid pattern for preschool children also compares favorably with the patterns derived from amino acid requirements of older children and adults.[11] Thus, it is the most widely used standard to assess the amino acid quality of a protein food, except for young infants.

As Table 2 shows, the current (1985) pattern of amino acid requirements has lower values for isoleucine, methionine + cystine, threonine, and valine than the provisional pattern proposed in 1973. Therefore, it has had important effects on the outcome of evaluating foods and diets in terms of their protein quality. The fact that the methionine + cystine requirements are less than what was previously estimated indicates that legumes, and particularly soy, can play an important nutritional role either as the main proteins in the diet or to supplement and improve the amino acid content and balance of foods and meals.

IV. NUTRITIONAL QUALITY OF SOY PROTEINS

Isolated soy proteins are relatively new protein sources. They were initially used in the 1960s in liquid formulas for babies intolerant to milk. Methionine was added to fortify the formulas in order to satisfy the estimated requirements of infants.

In view of the new knowledge about methionine + cystine requirements and on the amino acid composition of soy protein (Table 2), several studies were done to determine the nutritional quality for preschool children of isolated soy protein without amino acid fortification.

V. EXPERIMENTAL DESIGN AND METHODS

Nitrogen (N) balance techniques were used to estimate the protein value of the isolated soy proteins.[12,13] Regression analysis was used to calculate regression coefficients between N intake or absorption and N retention of the test proteins, and were compared with those of milk as a reference protein. The regression coefficient of intake on retention provides an estimate which is similar to net protein utilization.[13] The regression coefficient of absorption on retention is similar to biological value.[13] The higher the regression coefficient, the better is the quality or value of the protein.[13,14] Apparent nitrogen digestibility was calculated and compared with that of milk protein as a reference. "True" digestibility was calculated from the data on apparent digestibility, assuming an obligatory fecal nitrogen loss of 20 mg/kg/d.[9]

Three separate studies with preschool-age children were carried out using three isolated soy proteins at INCAP.[10,15] An additional study was conducted at the Chilean Institute of Nutrition and Food Technology (INTA) with a fourth isolated soy protein.[16] The isolated soy proteins initially identified as Supro® 620, Supro® 710, and Purina® Protein 220 were evaluated at INCAP, and SPF-200 was tested at INTA. Equivalent products are now identified as Supro® 710, Supro® 1751, Supro® 1711 from Supro® 710; Supro® 500E, ProPlus 500F, Supro® 620, Supro® 1500, and Supro® 760 from Supro® 620; PP220 from Purina® Protein 220; and Supro® 200G, ProPlus 200F, ProPlus 200FC from SPF-200. The basic process of producing these products remains unchanged and they differ only in brand name, cation used to neutralize the protein (sodium, potassium), and nutrient fortification to meet specific food requirements or microbiological specifications.

In the first study,[15] nitrogen intake levels ranging from 320 to 120 mg/kg/d (equivalent to 2.0 to 0.75 g protein, based on N × 6.25) were tested in descending order at 9-d intervals. The first 5 d in each period were for stabilization at the new level of N intake, and metabolic balance studies were done in the last 4 d. All diets supplied 100 kcal/kg/d and were supplemented with vitamins and minerals to meet the requirements of preschool children. Subsequent studies[10,15] evaluated N intakes in the range of 200 to 80 mg N per kilogram per day (1.25 to 0.5 g soy protein, N × 6.25), since it was apparent that higher intakes of soy proteins largely exceeded requirements and could not be used in the calculation of regression coefficients.[15] Either a descending order of N intake[15] or a combination of ascending and descending designs[10] were used. In the study with Supro® 710,[15] the effect of choline supplementation was studied using soy protein formulas with or without the addition

TABLE 3
Composition of a Typical Liquid Diet to Supply 1.25 g Protein Per Kilogram Per Day

Component	Amount (g)
Isolated soy protein	1.44[a]
Cornstarch	1.50[a]
Sugar	15.00[a]
Cottonseed or peanut oil	3.26
Mineral mixture[b]	0.61
Water, to total of	80.00

[a] Changes in protein intake were accompanied by equivalent isoenergetic changes in cornstarch or sugar.

[b] Provides (in meg): K, 6; Na, 1; Ca, 0.5; Mg, 0.2; Cl, 6; PO_4, 0.5; CO_3, 1; SO_4, 0.2.

of 50 mg choline chloride daily, which is equivalent to the amount of free choline present in about 500 ml of cow milk.[17] This was done based on reports which suggest that, at least for other animal species, the need for choline may increase when methionine is the limiting amino acid.[18-20]

These studies were with children in the age range of 19 to 44 months. The children were housed in the INCAP Clinical Research Center under 24-h supervision. They had been originally admitted to the Center due to malnutrition, but they had recovered completely at least 1 month prior to the initiation of the studies with soy proteins. The children were given the test meals as liquid formulas using refined diets (Table 3). Changes in protein concentration were compensated with isoenergetic changes in carbohydrates. When milk protein was evaluated, skim milk replaced isolated soy protein and sugar in isonitrogenous and isoenergetic amounts. The diets were fed as five identical daily meals. Vitamin and mineral supplements were given daily to meet the child's requirements.

The study of Egana et al.[16] used the same basic procedures with some slight modifications. The children were slightly older (31 to 62 months) and were transferred from a local orphanage to the INTA Clinical Center. The diets were a combination of liquid and solid food prepared from refined ingredients similar to those used in the INCAP studies. The nitrogen intake levels ranged from 240 to 120 mg/kg/d to provide protein levels of 1.5 to 0.75 g/kg/d (N × 6.25). The feeding period was 8 d for each protein, with 4 d for stabilization and the last 4 to measure nitrogen balance.

VI. RESULTS AND DISCUSSION

The overall acceptance of the isolated soy protein was good in all the studies. No unusual clinical changes were observed in the children by the medical and paramedical staffs nor in the biochemical parameters measured during the course of the studies.

The range of nitrogen intakes varied from 80 to 320 mg/kg/d, which gave considerable differences in apparent digestibility values.[10,15,16] However, when the apparent digestibilities were corrected for obligatory fecal nitrogen losses of 20 mg/kg/d,[9,21] the differences in digestibility due to the different intakes were largely eliminated.

The estimated "true" digestibilities of the four isolated soy proteins are summarized in Table 4. They were similar to the digestibility of milk proteins, with average values ranging from 92 to 96%. There was no effect of the level of protein intake on the estimated "true" digestibility, although there were small variations in the individual data points. The equivalency in "true" digestibility was also seen when milk and isolated soy proteins were compared in the same study.

TABLE 4
Estimated "True" Digestibility (Percent of Intake) of Isolated Soybean Protein and Milk Protein at Various Levels of Nitrogen Intake

N intake (mg/kg/d)	Isolated soy protein					Cow milk		
	Supro® 620[15]	Supro® 710[15]		PP220[a]	SPF200[b]	Dry skim milk[a]	Dry full-fat milk[b]	Various lots[c]
		+ Choline	No choline					
320	93.2	—	—		—		—	90.2
280	95.1	—	—		—		—	90.1
240	95.3	—	—		94.5		92.8	90.3
200	98.0	95.0	96.0		93.2		88.7	93.0
160	97.0	94.5	97.5		93.4		88.9	90.5
120	98.6	95.7	96.7		88.3		91.4	—
80	—	99.0	93.0		—		—	94.0
Mean, all	96	96	96	94	92	94	90	91
Mean, 80 to 200 mg	98	96	96	94	92	94	90	93

[a] Combined data for ascending and descending study designs.[10]
[b] Recalculated from Egana et al.[16] using 20 mg N per kilogram per day as obligatory fecal N loss.
[c] Summarized by Torun[15] from other INCAP studies.[22-27]

TABLE 5
Regression Equations of Apparent Nitrogen Retention (Y) on Nitrogen Intake (X), and Estimates of Safe levels of Protein Intake

Protein source	Regression equation[a] $Y = a + bX$	N Intake to retain 36 mg/kg/day[b]	Safe level of protein intake, g/kg/day[c]	Ref.
Supro® 620	−56.3 + 0.65X (41.8, 0.20)	142	1.11	15
Supro® 710	−34.6 + 0.59X (17.1, 0.13)	120	0.93	15
PP220	−46.0 + 0.58X (15.0, 0.09)	141	1.10	10
SPF-200	−54.0 + 0.49X (9.4, 0.10)	184	1.43	16
Milk	−45.0 + 0.70X (22.0, 0.12)	116	0.90	10
Milk	−33.6 + 0.51X (16.2, 0.08)	136	1.07	16
Milk	−53.8 + 0.64X	140	1.10	22—27

[a] Mean of individual data on the intercepts (a) and regression coefficients (b), with corresponding standard deviations in parentheses.
[b] 26 and 10 mg N per kilogram per day for growth and miscellaneous losses, respectively.[11]
[c] N × 6.25 plus twice the coefficient of variation of 12.5%.[11]

Table 4 also shows that the addition of choline to the soy protein had no effect on digestibility.

Table 5 shows the regression equations of apparent nitrogen retention (i.e., without including sweat and miscellaneous losses) on nitrogen intake for the four isolated soy proteins and cow milk from different studies. The amounts of dietary nitrogen required to satisfy the needs of these preschool children were calculated from the equations, in order to allow a retention of 26 mg N per kilogram per day for growth of children 2 to 4 years old and 10 mg N per kilogram per day for sweat and miscellaneous losses.[11] Those amounts were then converted to protein, using a factor of 6.25, and 25% more was added corresponding to twice the population coefficients of variation to calculate the safe levels of protein intake.[11]

The safe levels of intake of the isolated soy proteins coincided with or were lower than current protein recommendations for children of this age group,[11] except for the Chilean study, which suggested the need of 0.3 g more soy than milk protein per kilogram per day.[16] On the average, the safe level of intake reported for milk by these investigators (1.02 g/kg/d) was 11% lower than the average value of 1.14 g/kg/d calculated for the four soy protein isolates.

Comparing the regression coefficients shown in Table 5 for the soy proteins and milk, the quality of the former ranged from 83 to 96% (average: 93%) in relation to milk. As was the case for digestibility, the addition of choline did not influence nitrogen retention, protein quality estimated from the regression equations, nor the safe level of soy protein intake.[15]

Table 6 shows the plasma protein concentrations after 9 d with each level of protein intake in the Guatemalan studies. Decreases were observed with intakes of 0.75 g soy or 0.5 g milk protein per kilogram per day. No differences with either soy or milk were observed between 1.5 and 0.75 g/kg/d in the Chilean study.[16,28] This suggests that, at least for the short duration of each level of intake, there were no deleterious effects with the soy isolates eaten at or below the recommended safe levels of protein intake.

VII. CONCLUSIONS

These studies showed that isolated soy proteins have a high digestibility. Their nutritional quality, although somewhat lower than that of milk proteins, is good enough to sustain adequate nitrogen balance when provided in the amounts recommended internationally for children of preschool age. These results and their essential amino acid pattern support the use of isolated soy proteins as the main protein in the diet and for blending with other foods to increase the protein supply of the diet and/or improve the quality of protein sources that are limited by their content of certain essential amino acids.

TABLE 6
Plasma Protein Concentrations (g/dl) with Different Levels of Protein Intake (mean / SD)

Protein (or N) intake (g or mg/kg/d)	Supro® 620	Supro® 710 + choline	Supro® 710	PP220		Milk	
				Descending	Ascending	Descending	Ascending
2.00 (320)	6.9 / 0.2			6.9 + 0.2		6.9 + 0.2	
1.75 (280)	7.0 / 0.3				7.0 / 0.4		6.9 / .03
1.50 (240)	6.9 / 0.3						
1.25 (200)	6.8 / 0.3	6.5 / 0.3	6.8 / 0.4	6.8 / 0.4	6.8 / 0.5	6.8 / 0.3	6.7 / 0.2
1.00 (160)	6.6 / 0.4	6.4 / 0.3	6.4 / 0.2	6.6 / 0.2	6.3 / 0.6	6.9 / 0.5	6.6 / 0.7
0.75 (120)		6.3 / 0.5	6.3 / 0.3[b]	6.3 / 0.2[b,c]	6.2 / 0.4[b]	6.6 / 0.3	6.8 / 1.0
0.50 (80)	6.4 / 0.5[a]	6.0 / 0.4[a]	5.9 / 0.3[a]	6.2 / 0.6[b]	6.2 / 0.4[b,c]	6.4 / 0.4[b]	6.5 / 0.3[b,c]

[a] Less than other levels of intake in same column, $p < 0.05$ (Student's paired t)

[b] Less than protein intake of 1.25 g/kg/d, $p < 0.01$ (Student's paired t)

[c] Different from preceding protein level. In the "ascending" design, 2 g/kg/d preceded the intake of 0.50 g/kg/d.

REFERENCES

1. **Behar, M.,** The story of Incaparina; utilization of available sources of vegetable protein for human feeding, *J. Am. Med. Women's Assoc.,* 18, 384, 1963.

2. **Steinke, F. H., Prescher, E. E., and Hopkins, D. T.,** Nutritional evaluation (PER) of isolated soy protein and combinations of food proteins, *J. Food Sci.,* 45, 323, 1980.

3. **Nakagawa, I., Takahashi, T., Suzuki, T., and Kobayashi, K.,** Amino acid requirements of children: minimal needs of tryptophan, arginine and histidine based on nitrogen balance method, *J. Nutr.,* 80, 305, 1963.

4. **Nakagawa, I., Takahashi, T., Suzuki, T., and Kobayashi, K.,** Amino acid requirements of children: nitrogen balance at minimal level of essential amino acids, *J. Nutr.,* 83, 115, 1964.

5. *Energy and Protein Requirements,* FAO/WHO, Tech. Rep. Ser. No. 522, World Health Organization, Geneva, 1973.

6. **Beaton, G. H., Calloway, D. H., and Waterlow, J.,** Protein and energy requirements, a joint FAO/WHO memorandum, *Bull. W.H.O.,* 57, 65, 1979.

7. **Torun, B., Pineda, O., Viteri, F. E., and Arroyave, G.,** Use of amino acid composition data to predict protein nutritive value for children with specific reference to new estimates of their essential amino acid requirements, in *Protein Quality in Humans: Assessment and In Vitro Estimation,* Bodwell, C. E., Adkins, J. S., and Hopkins, D. T., Eds., AVI Publishing, Westport, CT, 1981, 374.

8. **Pineda, O., Torun, B., Viteri, F. E., and Arroyave, G.,** Protein quality in relation to estimates of essential amino acid requirements, in *Protein Quality in Humans: Assessment and In Vitro Estimation,* Bodwell, C. E., Adkins, J. S., and Hopkins, D. T., Eds., AVI Publishing, Westport, CT, 1981, 29.

9. **Torun, B., Cabrera-Santiago, M. I., and Viteri, F. E.,** Protein requirements of preschool children: obligatory nitrogen losses and nitrogen balance measurements using cow's milk, *Arch. Latinoam. Nutr.,* 31, 571, 1981.

10. **Torun, B., Cabrera-Santiago, M. I., and Viteri, F. E.,** Protein requirements of preschool children: milk and soybean protein isolate, in *Protein Energy Requirements of Developing Countries: Evaluation of New Data,* Torun, B., Young, V. R., and Rand, W. M., Eds., Food Nutr. Bull. Suppl. 5, United Nations University, Tokyo, 1981, 182.

11. Energy and Protein Requirements, FAO/WHO/UNU, Report of a Joint FAO/WHO/UNU Expert Consultation, Tech. Rep. Ser. No. 724, World Health Organization, Geneva, 1985.

12. **Allison, J. B.,** Biological Evaluation of Proteins, *Physiol. Rev.,* 35, 664, 1955.

13. *Nutritional Evaluation of Protein Foods,* Pellett, P. L. and Young, V. R., Eds., Food Nutr. Bull. Suppl. 4, United Nations University, Tokyo, 1980.

14. **Viteri, F. E. and Bressani, R.,** The quality of new sources of protein and their suitability for weanlings and young children, *Bull. W.H.O.,* 46, 827, 1972.

15. **Torun, B.,** Nutritional quality of soybean protein isolates: studies in children of preschool age, in *Soy Protein and Human Nutrition,* Wilcke, H. L., Hopkins, D. T., and Waggle, D. H., Eds., Academic Press, New York, 1979, 101.

16. **Egana, J. I., Fuentes, A. Steinke, F. H., and Uauy, R.,** Protein quality comparison of a new isolated soy protein and milk in Chilean preschool children, *Nutr. Res.,* 3, 195, 1983.

17. **Macey, I. C., Kelly, H. J., and Sloan, R. E.,** The Composition of Milks, Publ. No. 254, National Academy of Sciences, Washington, D.C., 1953.

18. **Du Vigneaud, V., Cohn, M., Chandler, J. P., Schenck, J. R., and Simmonds, S.,** Utilization of methyl group of methionine in biological synthesis of choline and creatine, *J. Biol. Chem.,* 140, 625, 1941.

19. **Treadwell, C. R.,** Growth and lipotropism: effects of dietary methionine, cystine and choline in young white rats, *J. Biol. Chem.,* 176, 1141, 1948.

20. **Mudd. S. H. and Poole, J. R.,** Labile methyl balances for normal humans on various dietary regimen, *Metab. Clin. Exp.,* 24, 721, 1975.

21. **Torun, B. and Viteri, F. E.,** Obligatory nitrogen losses and factorial calculations of protein requirements of preschool children, in *Protein Energy Requirements of Developing Countries: Evaluation of New Data,* Torun, B., Young, V. R., and Rand, W. M., Eds., Food Nutr. Bull., Suppl. 5, United Nations University, Tokyo, 1981, 159.

22. **Scrimshaw, N. S., Bressani, R., Behar, M., and Viteri, F.,** Supplementation of cereal proteins with amino acids. I. Effect of amino acid supplementation of corn-masa at high levels of protein intake on the nitrogen retention of young children, *J. Nutr.,* 66, 485, 1958.

23. **Bressani, R., Scrimshaw, N. S., Behar, M., and Viteri, F.,** Supplementation of cereal proteins with amino acids. II. Effect of amino acid supplementation of corn-masa at intermediate levels of protein intake on the nitrogen retention of young children, *J. Nutr.,* 66, 501, 1958.

24. **Bressani, R., Wilson, D., Chung, M., Behar, M., and Scrimshaw, N. S.,** Supplementation of cereal proteins with amino acids. V. Effect of supplementing lime-treated corn with different levels of lysine, tryptophan and isoleucine on the nitrogen retention of young children, *J. Nutr.,* 80, 80, 1963.

25. **Bressani, R., Viteri, F., Elias, L. G., Zaghi, S. de, Alvarado, J., and Odell, A. D.,** Protein quality of a soybean protein textured food in experimental animals and children, *J. Nutr.,* 93, 349, 1967.

26. **Bressani, R., Alvarado, J., and Viteri, F. E.,** Evaluacion, en ninos, de la calidad de la proteina delmaiz opaco-2, *Arch. Latinoam. Nutr.,* 19, 129, 1969.

27. **Bressani, R., Viteri, F. E., Wilson, D., and Alvarado, J.,** The quality of various animal and vegetable proteins with a note on the endogenous and fecal nitrogen excretion of children, *Arch. Latinoam. Nutr.,* 22, 227, 1972.

28. **Egana, J. I., Fuentes, A., and Uauy, R.,** Protein needs of Chilean preschool children fed milk and soy protein isolate diets, in *Protein Energy Requirement Studies in Developing Countries: Results of International Research,* Rand, W. M., Uauy, R., and Scrimshaw, N. S., Eds., Food Nutr. Bull., Suppl. 10, United Nations University, 1981, 249.

Chapter 11

CLINICAL EVALUATION OF PARTIALLY HYDROLYZED SOY PROTEIN AND CASEIN FOR BIOLOGICAL VALUE

Taisya A. Yatsyshina, Vadim G. Vysotsky, Anna M. Safronova, and
Alexander S. Vitollo

Providing the population with food protein in accordance with the recommended norm of physiological requirements[1] and rational amounts of major types of protein is an important objective for developing and industrially developed countries in the world. In the developing countries, the problem is mainly associated with insufficient production of protein food products. In the developed countries, dietary problems are primarily due to overconsumption or specific health conditions which require the production of specialized foods. The last aspect has drawn increasing attention due to the practical requirement of correcting human food status at various stages of current diseases, stress conditions, and the need to replace breast milk.

The majority of specialized foods for these purposes are developed based on the use of cow milk and soy proteins as well as their hydrolysates.[2] Studies conducted with laboratory animals showed[3] that the biological value of soy proteins is somewhat less than that of milk proteins, and thus show a necessity to enrich them with sulfur-containing amino acids. Similar results were obtained while testing such proteins in studies involving babies,[4] pre-school children,[5] and adults.[6] These studies determined that deficiency of sulfur-containing amino acids in soy proteins plays a more important role in the nutrition of children than of adults. The practical implementation of the above is to enrich soy proteins with methionine in breast milk replacements[7] to increase the biological value of the protein.

The use of hydrolysates of milk and soy proteins in the formulations of products for enteral and children's nutrition is due to the need for increased digestibility of protein in the gastrointestinal tract as well as to provide amino acid compositions in specialized products designed for patients with specific types of pathologies.[2,7]

Considering the above, as well as previously identified[3] differences in protein biological value figures obtained in experiments with laboratory rats and in human observations, we conducted two series of studies concerning comparative evaluation of the quality of slightly enzymatically modified PP710 soy protein (Ralston Purina Company, U.S.) and casein subjected to mild enzymatic hydrolysis (mean content of free amine nitrogen is about 5%).

In the first series of studies the biological value of the modified casein and soy proteins was studied in a comparative aspect. The experiment was conducted on 20 growing male Wistar rats with an initial body weight of 53.3 $\pm$ 0.6 g, subdivided according to protein type into two groups of ten animals each. During the 28-d experiment, the rats were housed individually in cages with unlimited consumption of water and semisynthetic diet containing 440 kcal for 100 g of dry mix. The diet included by weight 11.5% of sunflower oil, 4% mineral mix,[8] 1% water and fat-soluble vitamins,[8] and 9.5% of modified casein (Group 1) or 9.7% PP710 soy protein (Group 2). The remainder was maize starch. During the experiment the animals were fed daily and the amount of feed consumed was monitored. The protein efficiency ratio was calculated based on the results obtained.[8]

A second series of comparative studies were conducted with 11 volunteers (7 males and 4 females) under controlled hospital conditions to evaluate tolerance and digestibility of the above proteins as well as their effect on some blood values, urine excretion of final nitrogen metabolism products, and nitrogen balance status. The initial physical characteristics of the

TABLE 1
Physical Characteristics of Volunteers

Subject no.	Sex	Age (years)	Body weight (kg)	Height (cm)
1	M	27	86.0	188
2	M	43	62.0	170
3	M	39	86.0	182
4	M	30	86.0	169
5	M	28	86.0	184
6	M	30	86.0	164
7	M	28	86.0	168
Mean ± SE		35.0 ± 2.4	72.7 ± 4.2	174.3 ± 3.0
8	F	48	86.0	165
9	F	27	86.0	173
10	F	31	86.0	165
11	F	47	86.0	160
Mean ± SE		38.2 ± 5.9	64.2 ± 2.2	165.6 ± 3.7

TABLE 2
Chemical Composition of Food Rations

Period of examination	Proteins (g/d) N	Proteins (g/d) N × 6, 25	Fats (g/d)	Carbohydrates (g/d)	Caloric value (kcal/d)
I. (Soya)	14.6	91.3	100.0	370.0	2745
II. (Casein)	14.7	91.9	100.0	370.0	2748

Note: Fats: one third butter, two thirds sunflower oil. Carbohydrates: dextrin-maltose and saccharose, 58 and 42%, respectively. Daily: 1 dragée of multivitamins, calcium gluconate 0.5 g, and dog rose broth 200 ml.

volunteers are shown in Table 1. Prior to the studies, the volunteers were on uncontrolled conventional diets. The study consisted of two successive 14-d periods. Supro® 710 isolated soy protein was tested in the first period and modified casein in the second period. During the study, volunteers received similar types of diet consisting of the following set of products (grams per day): Supro® 710 soy protein, 85 (by protein) or modified casein, 85 (by protein); sunflower oil, 10; melted butter, 60; sugar, 100; dextrin-maltose, 5; starch, 16; "Saco" starch cereal, 120; cabbage, 500; carrots, 100; dry onions, 50; fresh apples, 500; raisins, 50; "Borzhomi" mineral water, 500; coffee, 5; tea, 3; fruit juice, 200; dog rose extract, 200; calcium gluconate, 0.5; and one "Undevit" multivitamin tablet.

The above products were consumed as protein drinks with water and fruit juices, vegetable salad with sunflower oil, "Saco" porridge with melted butter, tea and coffee with sugar, apple juice gel, and dog rose extract with sugar. Meals were consumed three times daily and divided by calories as breakfast and supper, 30% of daily calories each, and for lunch 40% of calories. During the day they did easy physical and mental work that did not exceed 2800 kcal/d if calculated by the chronometry method.[8] The chemical composition of daily diets is shown in Table 2. For the last 3 d of each 14-d period, male volunteers were measured for daily urine excretion of total nitrogen, urea, creatinine amine nitrogen, and ammonia. Fecal nitrogen losses were analyzed as the total for 3 d. The content of total nitrogen in urine, feces, and diets of volunteers as well as in feed for rats was determined by the Kjeldahl method using the Kjetech-Auto-1039 Analyzer by Techator (Sweden). Nitrogen balance status was calculated based on these results.[8] Urea and creatinine content in daily urine were determined using reagent kits by Lachema (Czechoslovakia), amine

TABLE 3
Biological Value of Soy Protein and Casein Hydrolyzates, Mean ± Standard Error

Protein source	Protein in diet (g/100 g)	Weight gain (g/d)	Feed consumption (g/100)	Protein consumed	Protein efficiency ratio
Soy protein	9.75	2.65 ± 0.10	9.76 ± 0.22	0.87[a] ± 0.02	3.03[a] ± 0.10
Modified casein	9.56	4.04 ± 0.24	11.88 ± 0.46	1.09[a] ± 0.04	3.53[a] ± 0.11

[a] $p < 0.05$.

TABLE 4
Body Weight of Volunteers

Subject no.	Sex	Initial	Day 14, period 1 Difference Actual vs. initial		Day 14, period 2 Difference Actual vs. initial	
1	M	86.0	83.5	−2.5	82.9	−3.1
2	M	62.0	61.7	−0.3	60.0	−2.0
3	M	80.7	77.0	−3.7	77.4	−3.3
4	M	61.2	61.7	+0.5	62.3	+1.1
5	M	76.0	74.0	−2.0	72.0	−4.0
6	M	81.0	78.5	−2.5	77.5	−3.5
7	M	60.0	58.8	−1.2	58.3	−1.7
Mean ± SE		72.7	70.7	−1.7	70.1	−2.4
		3.9	3.7	0.6	3.7	1.4
8	F	64.6	62.3	−2.3	62.5	−2.1
9	F	61.0	60.8	−0.2	61.2	+0.2
10	F	69.0	66.8	−2.2	65.3	−3.7
11	F	62.2	61.2	−1.0	61.5	−0.7
Mean ± SE		64.2	62.8	−1.4	62.6	−1.6
		2.2	1.7	0.6	1.2	1.1

nitrogen by the Pope and Stevens method,[9] and ammonia by the Convay microdiffusion method.[9] These assays were not performed with the female volunteers due to variances in menstrual cycles. Prior to the beginning of the studies and on day 13 of each period, 10 ml of veinous blood was taken from each volunteer. The blood serum was analyzed for the following: total protein, albumin, urea, uric acid, creatinine, cholesterol, glucose, bilirubin, calcium, phosphorus, iron, and hemoglobin, as well as alanine transferase and asparagine transferase. Lactate dehydrogenase, alkali phosphatase, and amylase were determined with Yanako (Japan) reagent kits and using the blood biochemistry analyzer Raba-Super, sodium and potassium serum concentration were determined using an AA spectrophotometer by Hitachi (Japan). Leukocyte and lymphocyte counts were determined in peripheral blood using conventional methods. All volunteers were examined daily, monitoring their well being, body weight, pulse frequency, and arterial pressure. Study data were statistically processed,[10] obtaining M ± SE values (mean + standard error).

The experimental data obtained in the first series of study are shown in Table 3. It follows from Table 3 that, as expected, the soy protein biological value was lower than that of modified casein and was 86% of the latter with the growing rat. In the second study, the well being of volunteers remained satisfactory during the entire period. Subjectively, some of them in the beginning of the first period (from 1 to 3 d) felt a slight discomfort as a minor heaviness in the abdomen. General medical standard indices were within the normal variation limits. Table 4 shows results of the volunteers' body weight determinations as related to test periods. Some body weight loss was observed between the initial and the final mean values in males and females, but this did not differ statistically ($p < 0.05$).

TABLE 5
Indices Used In Examining the Blood (Mean ± SE)

Index		Normal fluctuation limits	Values by period		
			Initial	Period I (soya)	Period II (casein)
Overall protein	g/l	65—85	60.3 ± 2.7	71.7 ± 2.9	81.4 ± 1.5
Albumins	g/l	37—52	34.9 ± 1.3	39.9 ± 1.5	43.1 ± 0.8
Urea	mmol/l	2.5—8.3	4.5 ± 0.2	4.3 ± 0.3	4.5 ± 0.2
Uric acid	μmol/l	119—238	184.1 ± 14.7	242.0 ± 30.3	203.5 ± 9.7
Creatinine	μmol/l	44—155	74.6 ± 2.9	96.9 ± 3.5	107.2 ± 4.3
Cholesterol	mg%	150—240	194.0 ± 20.0	188.8 ± 22.4	180.4 ± 19.5
Glucose	mmol/l	3.5—5.6	3.7 ± 0.1	4.5 ± 0.3	4.5 ± 0.1
Bilirubin	μmol/l	20.5	7.9 ± 0.7	8.5 ± 1.1	6.8 ± 0.6
ALT[a]	U/l	5—40	32.5 ± 2.6	30.2 ± 2.8	32.9 ± 3.5
ALT[b]	U/l	5—40	31.3 ± 3.3	30.1 ± 3.1	31.8 ± 2.7
LDH	U/l	300—500	130.5 ± 16.2	196.9 ± 17.9	186.6 ± 19.1
Alkaline phosphatase	U/l	10—50	3.82 ± 3.8	46.2 ± 6.0	34.2 ± 2.1
Amylase	U/l	0.8—2.1	1.2 ± 0.2	1.4 ± 0.2	1.3 ± 0.2
Sodium	mmol/l	145—155	145.2 ± 2.8	145.0 ± 2.8	152.8 ± 1.7
Potassium	mmol/l	4.0—6.0	4.4 ± 0.1	4.5 ± 0.2	4.8 ± 0.2
Calcium	mg/100 ml	9.0—11.0	9.9 ± 0.1	9.3 ± 0.1	9.7 ± 0.1
Phosphorus	mg/100 ml	2.0—4.5	3.8 ± 0.1	2.8 ± 0.2	3.6 ± 0.2
Iron	μmol/l	10.7—26.0	15.9 ± 1.6	25.5 ± 1.1	18.2 ± 1.0
Hemoglobin	g%	12—17	13.9 ± 3.2	14.9 ± 2.0	15.1 ± 1.6
Leukocytes	in 1 mm³	6000—8000	6555 ± 810	5909 ± 336	5259 ± 541
Lymphocytes	in 1 mm³	1500—3000	2385 ± 338	2002 ± 189	2042 ± 381

[a] Alanine aminotransferase.
[b] Aspartic glutamic transaminase.

Table 5 shows results of biochemical and morphological blood tests and shows that the protein source consumed by the volunteers did not affect the indices significantly, because the observed values were within the normal range.

Nitrogen balance and protein digestibility test results are shown in Table 6. Attention is drawn to the somewhat lower true digestibility of Supro® 710 soy protein as compared to that of modified casein, but the difference was not statistically different ($p < 0.05$). A tendency towards ($p < 0.05$) an increase in nitrogen balance value in the second period was found, and this to some extent confirmed that modified casein has a higher biological value that was established in the experiments with rats. At the same time, apparent nitrogen equilibrium in both periods shows qualitative adequacy of these proteins with the consumption levels used.

Table 7 shows results of the final products of nitrogen metabolism excretion in urine. It follows that in the first period, i.e., while consuming PP710 soy protein, volunteers were observed to have slightly higher ($p < 0.05$) urinary excretion of urea, ammonia, and creatinine as compared to the second period. This to a certain degree confirms data on the difference in biological value of Supro® 710 soy protein and modified casein, as well as nitrogen balance status in the volunteers. Thus, as a result of these studies, a good tolerance of Supro® 710 soy protein and modified casein by humans was established and biological efficiency with the consumption levels used and the observation duration were adequate to maintain nitrogen equilibrium and values of biochemical indices studied within normal limits.

TABLE 6
Nitrogen Balance and Digestibility of Protein Hydrolysates

Period of examination	Subject no.	Excreted nitrogen (g/d) w/Urine	w/Feces	Nitrogen balance (+ g/d)	True digestibility (%)
I	1	13.1	1.4	+0.1	96
	2	11.5	2.2	+0.9	90
	3	13.8	2.1	−1.3	91
	4	9.6	1.7	+3.3	93
(Soya)	5	12.7	1.8	+0.1	93
	6	12.3	2.3	0	90
	7	11.8	3.2	−0.4	82
Average		12.1 ± 0.6	2.1 ± 0.1	+0.4 ± 0.7	90.7 ± 2.0
	1	12.0	1.5	+1.2	96
	2	12.8	2.0	−0.1	91
	3	13.9	0.9	−0.1	99
II	4	10.9	1.6	+2.2	94
	5	12.7	1.4	+0.6	96
(Casein)	6	13.5	1.1	+0.1	98
	7	11.9	1.3	+1.5	95
Average		12.5 ± 0.4	1.4 ± 0.2	+0.8 ± 0.4	95.6 ± 1.2

Note: The excretion of endogenic nitrogen with the feces was determined, calculated on 10.7 mg of nitrogen per kilogram of body weight per day. The quantity of consumed nitrogen measured 14.6 ± 0.2 g/d in period I, and 14.7 ± 0.1 g/d in period II.

TABLE 7
Excretion of End Products from Nitrogen Metabolism with the Urine

Products from nitrogen	Metabolism	Period I (soya)	Period II (casein)
Total nitrogen	g/d	12.1 ± 0.6	12.5 ± 0.4
Urea	g/d	19.6 ± 1.03	17.3 ± 0.5
Blood urea nitrogen	% of total nitrogen	75.7 ± 2.8	65.0 ± 3.6
Ammonia	g/d	0.45 ± 0.02	0.42 ± 0.02
Ammonia nitrogen	% of total nitrogen	3.05 ± 0.13	2.67 ± 0.14
Creatinine	g/d	2.01 ± 0.10	1.81 ± 0.14
Creatine nitrogen	% of total	6.23 ± 0.41	5.34 ± 0.39
Amino nitrogen	g/d	0.40 ± 0.03	0.40 ± 0.05
	% of total nitrogen	3.29 ± 0.15	3.19 ± 0.37

REFERENCES

1. Energy and Protein Requirements, Report of Joint FAO/WHO/UNU Expert Consultation, Tech. Rep. Ser. No. 724, World Health Organization, Geneva, 1985.
2. **Moses, N.,** Infant formula and enteral products to meet nutritional needs, in *Nutrition for Special Needs in Infancy, Protein Hydrolysates,* Lifshitz, F., Ed., Marcel Dekker, New York, 1985, 267.
3. **Bodwell, C. E.,** Human versus animal assays, in *Soy Protein and Human Nutrition,* Wilcke, H. L., Hopkins, D. T., and Waggle, D. H., Eds., Academic Press, New York, 1979, 331.
4. **Fomon, S. Y. and Ziegler, E. E.,** Soy protein isolates in infant feeding, in *Soy Protein and Human Nutrition,* Wilcke, H. L., Hopkins, D. T., and Waggle, D. H., Eds., Academic Press, New York, 1979, 79.

5. **Torun, B.,** Nutritional quality of soybean protein isolates: studies in children of preschool age, in *Soy Protein and Human Nutrition,* Wilcke, H. L., Hopkins, D T., and Waggle, D. H., Eds., Academic Press, New York, 1979, 101.

6. **Scrimshaw, N. S. and Young, V. R.,** Soy protein in adult human nutrition: a review with new data, in *Soy Protein and Human Nutrition,* Wilcke, H. L., Hopkins, D. T., and Waggle, D. H., Eds., Academic Press, New York, 1979, 121.

7. **Cook, D. A. and Sarett, H. P.,** Design of infant formulas for meeting normal and special needs, in *Pediatric Nutrition: Infant Feeding Deficiencies Diseases,* Lifshitz, F., Ed., Marcel Dekker, New York, 1982, 71.

8. Nutritional Evaluation of Protein Foods, UNU, Food and Nutrition Bulletin Suppl. 4, Pellet, P. L. and Young, V. R., Eds., United Nations University, Tokyo, 1980, 104.

9. **Rubin, V. T., Larskii, E. G., and Orlova, L. S.,** *Biochemical Methods of Research in Clinic,* Publishing House of Saratov University, 1980, 133.

10. **Kaminskii, L. S.,** *Statistic Method of Reading of Clinical and Laboratory Data,* Medicina, Leningrad, 1964, 21.

Chapter 12

PROTEIN NUTRITIONAL VALUE OF SOY PROTEINS IN ADULT HUMANS

Vernon R. Young

TABLE OF CONTENTS

I. INTRODUCTION

The role of the soybean as a traditional food item in diets of populations in Asia is well recognized. Advances in food technology resulting in the development of a variety of attractive edible soy products, including isolated soy protein, concentrates, and textured/ structured products, with characteristics that resemble other traditional foods, have now made available soy protein foods for use and enjoyment by populations in technically developed regions of the world. The expectations are that this increased usage of soy proteins will continue, particularly as more palatable and attractive soy protein-based foods are developed. There are now a variety of actual or potential food uses of soy within populations that earlier have not utilized soy to any commercially significant extent, and Table 1 lists some of these. Hence, in this chapter, the potential of soy to contribute significantly to the total protein intake and meet the physiological needs of adult humans will be reviewed. The nutritional properties of soy protein for infants and children are discussed elsewhere in this book (Chapters 8 and 10).

Emphasis will be given here specifically to the topic of protein nutrition and the nutritional value of soy protein-containing foods. The discussion will include a consideration of a series of major studies on the nutritional value of isolated soy protein in adults carried out in our laboratories at the Massachusetts Institute of Technology (MIT) and conducted under the major direction and leadership of my colleague, Professor Nevin S. Scrimshaw.

II. PROTEIN QUALITY EVALUATION IN ADULT HUMANS

Dietary proteins are needed for a variety of metabolic purposes, including replacement of tissue and organ proteins due to continuous metabolic losses, the formation and net deposition of protein in new tissues during development, pregnancy, for lactation, and replenishment of depleted tissues due to preceding pathological conditions. These needs are met specifically by the indispensable (essential) amino acids and nonspecific nitrogen sources (largely the dispensable amino acids) that comprise dietary proteins.

It is usually accepted that the concentration and availability of individual essential amino acids are the major factors in determining the nutritive value or quality of a food protein source.[1] Also, the extent to which a given source of food protein is capable of supporting an adequate nutritional state will depend on the physiological requirements of the individual for essential amino acids, as well as for total nitrogen. These principles are the basis for the procedures used for assessing the nutritional quality of a food protein, including the amino acid scoring approach (see Chapter 2).

The amino acid requirements of humans also have been discussed in an earlier chapter in this book (Chapter 2), and there it was suggested that the essential amino acids of the adult are considerably higher than those proposed by the most recent Food and Agriculture Organization/World Health Organization/United Nations University (FAO/WHO/UNU) Expert Consultation.[1] Recent recommendations made by us for the amino acid requirement in adults leads to the tentative conclusion that they are probably close to that of the 2- to 5-year-old child (Table 2), when expressed in relation to the protein need. As can be seen from Table 2, the amino acid pattern recommended by Young and Pellett[2] is similar to that for the young child except for being lower in lysine and threonine.

Perhaps because of the lack of a broad consensus on the acceptability of amino acid turnover and flux methodology for assessing amino acid requirements (see Chapter 2), a recent FAO/WHO Expert Consultation on Protein Quality Evaluation[3] adopted the amino acid requirement pattern for the 2- to 5-year-old child (Table 2) as that to be used for assessing protein nutritional quality by the amino acid scoring procedure for all ages.

Comparing the amino acid pattern of soy proteins (Table 3) with that of this reference

TABLE 1
Some Food Uses of Soy in Relation to Human Protein Nutrition

Use	Population group
Alternative to milk-based formulas	Infants
Milk-free foods	Infants and children
Vegetable protein mixtures	Preschool and school-age children
Protein-enriched drinks	Children/adults
Food analogs, substitutes, and extenders	All ages
Traditional food items	All ages

TABLE 2
Amino Acid Requirement Patterns for the Young Child and Adult as Proposed by FAO/WHO/UNU[1] and the Adult Pattern Proposed by Young and Pellett[2]

	FAO/WHO/UNU[1]		Young and Pellett[2]
Amino acid	2- to 5-Yr. old child	Adult	Adult
Isoleucine	28	13	38
Leucine	66	19	65
Lysine	58	16	50
Sulfur amino acids	25	17	25
Aromatic amino acids	63	19	65
Threonine	34	9	25
Tryptophan	11	5	10
Valine	35	13	35
TOTAL	320	111	313

TABLE 3
Comparison of the Amino Acid Concentration in Various Soy Protein Products with the 1985 FAO/WHO/UNU[1] Amino Acid Scoring, or Requirement, Pattern for the 2- to 5-Year-Old Child

	FAO/WHO/UNU 2- to 5-Year-old pattern	Soy protein product			
Amino acid		SPI[a]	SPC[b]	SPC[c]	SPC[d]
Isoleucine	28	49	49	50	52
Leucine	66	82	84	79	80
Lysine	58	64	62	64	64
Methionine and cystine (SAA)	25	26	28	30	30
Phenylalanine and tyrosine (AAA)	63	92	89	93	93
Threonine	34	38	44	40	41
Tryptophan	11	14	16	13	13
Valine	35	48	50	53	54
TOTAL	320	413	422	422	427

Note: Values are milligrams of amino acid per gram of crude protein (N × 6.25).

a   Soy protein isolate.[4]
b   Soy protein concentrate.[5]
c   Soy protein concentrate.[6]
d   Soy protein concentrate.[6]

TABLE 4
Some Used or Potential Measures for Evaluation of
Dietary Protein Quality in Human Subjects

Growth
   Weight
   Height
   Lean body mass
      40K Whole-body counting
      Body density
      Body water (isotope dilution)
      Creatinine height index
Blood and serum constituents
   Proteins: albumin, enzymes
   Free amino acids
   Urea N
Metabolic balance
   Nitrogen excretion
   Sulfur excretion
Amino acid oxidation and flux

pattern indicates that the soy protein meets this pattern and should therefore also meet the pattern of amino acid requirements for the adult, as well. This suggests that at protein intakes needed to meet the total nitrogen requirement (with high-quality protein, such as egg, milk, or meat), the soy protein would similarly satisfy the essential amino acid needs of both the child and the adult.

In a major series of studies conducted at MIT, we have tested this prediction and evaluated the nutritional value of soy proteins in young adults. Our results, and those of others discussed below, confirm the high value of this protein source, which has become increasingly available and incorporated into foods that both meet human nutritional needs and are consistent with dietary recommendations and guidelines regarding health maintenance and disease prevention.[7]

Clinical methods for the evaluation of protein quality are based on the same principles applied to the corresponding animal assays, but often require specific modification for application to human subjects.[8-10] The major procedures and criteria used are shown in Table 4. These include investigations of growth and nitrogen balance, alone or in combination with biochemical analysis of serum proteins and amino acids, hemoglobin, blood urea nitrogen, and the urinary excretion of creatinine, sulfur compounds, and hydroxyproline. The majority of previous studies conducted in humans for determining the nutritional value of soy protein have been based on measures of growth, N balance, and dietary N utilization in infants and children and on indices of N balance in adolescents and adults.

In nitrogen balance experiments with soy protein, the usual measurements made are depicted in Figure 1. Studies may include several different intake levels of test protein together with a reference protein source, such as a hen egg or human or cow milk. If multiple test levels of protein intake are studied, a determination is made of the relationship between nitrogen (protein) intake and nitrogen retention or N balance in each experiment and comparisons are made between the test and reference protein source.

Two different response criteria might be used to evaluate dietary protein quality from the N balance response curve[9] as depicted in Figure 1. The first is an estimation of the efficiency with which dietary nitrogen is utilized. This is assessed from the *slope* of the nitrogen balance response curve in the region of the submaintenance to near maintenance N intake level. The second estimate of protein quality can be derived from a comparison as to how well a given protein source meets the minimum requirement for total protein and amino acids, i.e., what minimum intake level of a given test protein is required to just

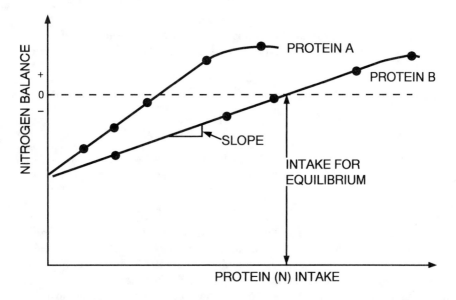

FIGURE 1.   Schematic depiction of N balance response criteria and two methods of estimation of dietary protein quality, in subjects receiving graded intakes of test protein within the sub-maintenance to near maintenance level of intake. The two methods involve comparison of slopes or intake for equilibrium.

maintain a zero nitrogen balance or body N equilibrium. In this case, the response criteria is the *intercept* of the N balance response curve with the line of zero N balance.

These two approaches for the comparative assessment of protein quality are related, but slightly different. However, with either, dietary protein quality can be studied from the relationship of the N balance responses to the test protein in comparison to those obtained with a suitable reference protein source.[9] Alternatively, some clinical studies have involved comparisons of N balance obtained with soy proteins and a reference protein determined only at a single level of test protein intake. This approach can provide valuable information, but care must be taken in evaluating the results of such studies because they may depend upon the particular test level of protein intake used for the comparison.[12]

Growth studies are based on the same general approaches as those used in N balance experiments, but usually must be conducted over longer periods of time. Because it is difficult to include a reference protein source and to control dietary intake as precisely as in short-term N balance experiments, interpretation of results obtained in growth studies might be more difficult. However, growth studies have provided much valuable information on the nutritional quality of soy protein, especially in relation to infant nutrition (see Chapter 8).

III. SOY PROTEIN QUALITY IN ADULT HUMANS

Many of the published studies on the nutritional quality of soy protein in adults have been reviewed,[13-15] together with results of an extensive series of nitrogen balance experiments in young adult male subjects who were studied at several different test protein levels in our own laboratories. The account that follows is essentially similar to that in a relatively recent review,[15] which may not be readily accessible and thus is presented here as a companion to the other chapters concerned specifically with the nutritional properties of proteins from the soybean.

Estimations of the nutritional value of soy protein isolates as derived from N balance data obtained in our original study,[4] using the approaches and criteria discussed above, are

TABLE 5
Summary of Results of Metabolic Nitrogen Balance Studies Conducted in Adults to Assess the Nutritional Quality of Soy Protein Isolates and Concentrates

Study and protein source	Experimental approach	Conclusions about value of SPI or SPC
Soy protein isolates (SPI)		
Young et al.[4]	Multiple N levels	~83% of value for egg protein
Wang et al.[16]	Multiple N levels	~80% of value for fish proteins
Kaneko et al.[17]	Multiple N levels	Equivalent to egg proteins
Scrimshaw et al.[18]	Multiple N levels	Equal to value for milk protein
Soy protein concentrate (SPC)		
Istfan et al.[5]	Multiple N levels	Equivalent to value for egg proteins

given in Table 5. The nutritional quality of the soy protein isolate tested (Supro® 620) in this experiment[4] was high for healthy adults, approximating better than 80% of the nutritional value of the reference, egg protein. The true digestibility (97%) of the soy isolate also was high and comparable to proteins in hen egg.[4] Of particular importance here is that the relative protein value of this soy isolate (Supro® 620, Protein Technologies International, St. Louis, MO) in adult humans is considerably higher than that which would be obtained from feeding studies in rapidly growing rats.[13] Clearly, this animal model would lead to an underestimate of the nutritional quality of the soy isolate for the human adult as well as for children.

There have been other short-term metabolic nitrogen balance studies by other investigators. Thus, also summarized in Table 5 are the findings by Wang et al.[16] and Kaneko et al.,[17] whose studies were conducted in Japanese adult males and females, and the results are comparable to those conducted in U.S. adults.[4,18] As can be seen, the nutritional value of soy protein isolates for both populations was found to be high and comparable to that of egg and/or fish proteins.

Since completing our initial series of experiments with isolated soy proteins,[4,18] we have carried out additional metabolic studies with soy protein concentrates (Table 5). Our purpose in extending our studies was to evaluate these alternative forms of soybean protein, and the results from a short-term metabolic nitrogen balance study with a soy protein concentrate[5] again revealed that the nutritional quality of this product was comparable to that of good-quality animal protein sources (see also Table 5). This conclusion parallels, therefore, that drawn above for well-processed isolated soy proteins.

IV. LONG-TERM METABOLIC STUDIES

A possible limitation of many of the human metabolic studies referred to above is that the experimental diet periods have been relatively brief, with most studies lasting less than 2 weeks. In view of the likely trend toward a continued increase in the use of soy in the diet of adult populations over time, we have conducted a series of controlled, longer-term metabolic experiments in healthy adults, to confirm and extend our findings from the short-term N balance studies.

Three such experiments have been completed.[6,19,20] The expense and complexity of such studies makes it unlikely that they will, or need, be replicated often, and so it is worth summarizing briefly the major observations emerging from them.

In the first of these studies,[19] with eight young adult men, we explored the longer-term capacity of a soy protein isolate (Supro® 710) to maintain parameters of protein nutritional status. The experiment consisted of an 84-d N balance study and the diet provided protein entirely from the soy isolate, at a level of 0.8 g protein per kilogram per day. The results for N balance, body weight changes, and whole-body potassium, as a measure of protein mass, showed maintenance of all of these parameters (Table 6).

TABLE 6
Summary of Major Findings from Three Long-Term Metabolic Studies to Evaluate the Nutritional Quality of a Soy Protein Isolate and Soy Protein Concentrates in Healthy Young Adult Males

Study	No. of subjects	Protein source and intake	Length of study (days)	Results
Young et al.[a]	8	SPI; 0.8	84	N balance maintained; blood chemistries normal; body composition maintained
Istfan et al.[b]	6	SPC; 0.8	82	N balance positive; blood chemistries no change; maintenance of performance in exercise tests
Beer et al.[c]	9	SPC; 0.8	77	N balance and basal metabolic rate and blood parameters maintained; good tolerance and acceptability
	8	SPC; 0.8		

[a] Protein source: SPI (Supro® 710), given at an intake of 0.8 g protein $kg^{-1}d^{-1}$ (Reference 19).
[b] Protein source: SPC (STAPRO 3200), given at an intake of 0.8 g protein $kg^{-1}d^{-1}$ (Reference 20).
[c] Protein source: SPC (Danprotex-H 40 or Danpro-S) given at an intake of 0.8 g protein $kg^{-1}d^{-1}$ (Reference 6).

The second long-term study[20] examined the capacity of a soy protein concentrate (STA-PRO-3200) to maintain parameters of protein nutritional status in six healthy young male MIT students. Following an initial 9-d period during which the subjects received an egg-protein formula diet supplying 1.5 g protein per kilogram per day, the soy protein was given at a level of 0.8 g protein (N × 6.25) per kilogram per day for 82 d. We found (Table 6) that nitrogen balances were slightly positive for all subjects, and the protein source was judged to support adequately the maintenance of protein nutritional status, confirmed by the absence of changes in relevant blood parameters and maintenance of performance in the exercise tests.

Finally, an elaborate long-term metabolic study[6] was carried out in 17 healthy young adult men. Here we wished to evaluate the tolerance to and protein nutritinal value of two commercially produced soy protein concentrates, Danpro-S and Danprotex-H 40 (produced by Aarhus Oliefabrik A/S, Denmark). The test protein intake again was 0.8 g/kg body weight per day, a level close to the 1985 FAO/WHO/UNU[1] safe intake for high-quality protein. The subjects remained healthy, and no problems of clinical or metabolic significance were experienced by any of the subjects during the experimental period. Body weight for each group remained constant, and body composition parameters were unaffected by the prolonged ingestion of the test diets (Table 6). Nitrogen (N) balances confirmed the good quality of the two soy protein concentrates tested. In addition, we included in this third, long-term experiment a number of immunological studies, and the results showed that there were no immunological disturbances brought about by giving these products as the entire source of dietary protein for a continuous period of about 3 months.

The results obtained from each of these three long-term metabolic studies confirm and extend the conclusions drawn from the short-term N balance experiments, summarized above, and they are consistent with the results of two "tolerance trials" that we conducted a number of years ago (see below). Clearly, for these soy protein products that we have evaluated in our laboratories, it can be stated with some assurance that they can be consumed over long periods of time as the sole source of dietary protein and are fully capable of maintaining protein nutritional status.

V. SUBSTITUTION OF SOY PROTEINS FOR ANIMAL OR OTHER PROTEIN SOURCES

The use of soy protein concentrates (about 70% protein) and of isolated soy proteins

TABLE 7

Summary of Major Findings from Metabolic Nitrogen Balance Studies, in Adults, to Assess Effects of Isolated Soy Proteins on Substitution for, or Combination with, Other Protein Sources

Study	Protein replaced or combined with	Interpretation of results
Wayler et al.[21]	Beef proteins	Soy protein isolate equivalent to that for beef and milk
Wang et al.[16]	Fish proteins	50:50 Combination of soy:fish equivalent to fish protein alone
Kaneko et al.[17]	Rice proteins	60:40 soy:rice combination equivalent to egg proteins

(90% protein), produced from defatted soybean flakes, in processed meats and in texturized products is a potentially important role for soy in the human diet. Therefore, assessments of the nutritive value of soy in combination with other proteins have been made in a number of studies with adult humans as the experimental model.

Again, the results of published studies concerned with the replacement of beef and fish proteins by soy protein products have been reviewed.[13,15] The available evidence indicates, as summarized in Table 7, that a broad variety of combinations of isolated soy protein with beef,[21] fish,[16] or rice[17] have a nutritional value essentially equivalent to that of high-quality animal protein sources. Thus, a significant or even complete replacement of either beef or fish proteins, for example, by these soy protein products would not result in any important change in the overall protein nutritional quality of the diet. These findings are entirely consistent with those reviewed above, involving an assessment of the quality of soy proteins when consumed as the sole source of indispensable amino acid and total nitrogen.

VI. SULFUR AMINO ACID CONTENT OF SOY PROTEIN IN RELATION TO HUMAN NUTRITION

Experiments with the young, rapidly growing rat clearly show that the sulfur amino acids (S-amino acids) are the most limiting essential amino acids in soy proteins and that their nutritional value is significantly enhanced by methionine supplementation.[13] It is therefore relevant to consider results of various human studies concerned with the nutritive value of soy protein in relation to the possible effects of methionine supplementation.

A number of the earlier studies on the effects of methionine supplementation on N balance or growth conducted in infants, children, and adolescents have been reviewed previously.[13,14] On the basis of these earlier studies in the young and of those carried out in adult subjects by Kies and Fox[22] and Zezulka and Calloway,[23] the utilization of the soy protein isolate was found to be improved at low intakes of test protein by methionine addition. However, when such tests are conducted at higher intakes (equivalent to about 0.6 g protein per kilogram) in adult subjects, at least, nitrogen balance is similar to that achieved with egg white nitrogen (about 0.4 g protein per kilogram per day) protein. These observations would be predicted from the amino acid scoring pattern referred to earlier (see Table 3), and the results of these investigations[22,23] indicate that soy proteins are fully capable of meeting the amino acid requirements of adult subjects without the need to supplement with methionine.

Additional investigations confirm and extend the conclusions drawn from these earlier investigations. For example, Fomon et al.[24] examined the consequences of methionine supplementation of soy protein incorporated into a formula intended for infant feeding. In a more recent series of studies designed to define the requirement of normal infants for sulfur-containing amino acids (methionine and cystine), Fomon et al.[25] also found a beneficial effect of methionine supplementation on nitrogen balance only when the formula provided

1.8 g isolated soy protein per 100 Kcal, but not at higher protein intakes tested, namely 2.2, 2.6, 2.8, and 3.0 g protein per 100 Kcal. These studies reveal that for infants, methionine or total sulfur amino acids may be limiting in soy protein isolates, depending upon the level of total protein intake. Studies in the infant primate (Cebus monkeys)[26] are consistent with these observations in human infants, showing that under specific circumstances the quality of a soy protein isolate is improved with methionine supplementation.

We[4] have also assessed the effects of methionine supplementation on the utilization of soy protein isolates in adult men. In the first study, methionine supplementation was given to about equal (supplementation at 1.1% of total protein) or somewhat exceed (1.6%) the level of total S-amino acid in the 1973 FAO/WHO Provisional Amino Acid Scoring Pattern.[27] This pattern provides 3.5% total S-amino acids or higher than that in the 1985 FAO/WHO/ UNU[1] amino acid requirement patterns for preschool-age children and older age groups. The effect of a lower level of methionine supplementation also was tested in our study,[4] since it was desirable to know the minimum level of supplementation necessary to achieve a maximal N balance response in adult subjects. The experiment was conducted at a level of test protein intake of 0.51 g soy protein per kilogram per day, chosen to be slightly lower than a maintenance level. Our results showing that methionine supplementation improved dietary N utilization in adults when soy was consumed as the sole source of protein at a *deficient* level of total N intake agree with those of Kies and Fox[22] and Zezulka and Calloway.[23] However, at the highest level (1.6%) of methionine supplementation studied, an apparent deterioration in overall nitrogen balance occurred. This observation is somewhat similar to those made in a series of earlier studies in young children, concerned with effects of amino acid supplementation of corn and wheat on nitrogen retention.[28-30] Not only did these studies show that the addition of methionine failed to improve N balance when corn or wheat provided the entire source of dietary protein in these studies, but addition of methionine to corn maize protein resulted in a *decrease* in nitrogen balance. From these observations, it is apparent that N utilization in children is determined in part by the dietary balance of essential amino acids, and a disproportion in the amino acid pattern can adversely affect dietary protein quality. Our study[4] also indicates that, at marginal or deficient levels of total N intake, healthy young men may react unfavorably to a relatively subtle imbalance among the essential amino acids.

A second study[4] was then conducted to confirm and evaluate further the effects of L-methionine supplementation of soy protein. Two levels of methionine supplementation, equivalent to 1.1 and 1.6% of total soy protein intake, were studied for protein intakes of 0.51 and 0.8 g protein $kg^{-1} \cdot d^{-1}$. The results confirmed that for the lower protein level, N balance improved with the 1.1% level of methionine supplementation in six of the eight subjects. However, there was a reduction in N balance with the 1.6% level of supplementation in these same subjects. Furthermore, when the test level of soy intake was equivalent to 0.8 g protein per kilogram per day, the two levels of methionine supplementation had no effect on N utilization. These findings in healthy U.S. adults are also supported by a series of investigations carried out in Japan, involving metabolic balance studies in adult Japanese men and women.[31-34]

Thus, it is clear that methionine supplementation is unnecessary in the case of the adult and that soy proteins, as isolates or concentrates, are excellent sole sources for meeting the nitrogen and all of the amino acid needs when consumed at physiologically significant intake levels of total protein. Methionine supplementation of soy-based infant formula may, however, be desirable, although it appears that the level of methionine addition required is modest.

VII. SUMMARY OF METABOLIC STUDIES

From these various investigations it can be concluded that the nutritional value of soy

TABLE 8
Design of a 6-Month Acceptability Study with a Soy Protein Isolate (Supro® 620)

Design
    Subjects: 40 Adult subjects (18 subjects experimental group; 22 subjects control group)
    Diets: Experimental — 40 g soy protein isolate (Supro® 620) daily
                           Control — 40 g DSM[a] powder daily
    Length of study: 6 months
    Approach: Clinical biochemical battery physical examinations reporting and appearance of clinical
                           reactions

[a]   Dried skim milk.

proteins in relation to meeting the amino acid and nitrogen needs of the adult is high. Their comparative quality closely approaches or is equal to that of proteins that have traditionally been used as reference sources, such as those of hen egg, cow milk, and meat. Furthermore, these metabolic data support the view, at the very least for the soy protein products which have been evaluated in this way, that the knowledge of their amino acid content provides an adequate basis for predicting this nutritional quality. Hence, we[35] believe that for routine purposes of food labeling, for example, it would now be entirely appropriate to use the protein digestibility-corrected amino acid scoring method, as proposed by FAO/WHO,[3] to estimate the nutritional value of well-processed soy protein products intended for use in human foods.

VIII. TOLERANCE AND ACCEPTABILITY OF SOY PROTEIN PRODUCTS

Even though soy protein is frequently used as a milk substitute for infants allergic to formulas based on cow milk, soy can also be a primary cause of immunological responses, and so questions have been raised about the possible consequences of a high intake of soy protein in man,[36] especially because the method of processing soy products can have an important effect on the antigenicity, or lack thereof, of these protein sources.[37,38] Thus, although allergic reactions are uncommon with traditionally processed soy-containing foods compared with many other common foods of animal and plant origin, the possible effects of new processing procedures should always be taken into account.

A comprehensive evaluation of the role of soy in human protein nutrition should therefore include studies on the acceptability of, and tolerance to, the long-term ingestion of the soy protein products that represent forms of preparation not previously experienced by man, or at levels of intake higher than those on which human experience has been accumulated. These studies must be conducted under close medical supervision and be supported by appropriate laboratory procedures. The results of long-term metbolic studies discussed above provide a strong basis for concluding that soy protein is well tolerated. These studies involved relatively few subjects, and so additionally we[42] also conducted two "acceptability" studies. In the first (Table 8), a total of 40 healthy adult subjects were selected and randomized into two groups: one consisted of 9 men and 9 women consuming daily 40 g of Supro® 620, and a double-blind group of 11 men and 11 women who consumed an equivalent amount of skim milk powder. The soy and milk powder were incorporated into fruit juices, and were consumed in addition to the free-choice diets of the subjects. One subject in the experimental group, a 21-year-old woman, developed milk gastrointestinal symptoms, nausea, and diarrhea approximately 2 h after consuming the material. Follow-up tests suggested that the symptoms in this subject were associated with the ingestion of Supro® 620 and commercial products containing it, but not with soy curd or textured vegetable protein (TVP) from another manufacturer. The reaction was mild and, with repeated testing, the sensitivity gradually disappeared before a specific cause could be identified.

TABLE 9
Design of a 2-Month Acceptability Study with a Soy Protein
Isolate (Supro® 710)

Design
   Subjects: 100 Adult subjects (50 experimental, 50 control)
   Diets: Experimental — 40 g soy protein isolate daily
                            Control — 40 g DSM[a] powder daily
   Length of study: 2 months
   Approach: Clinical biochemical battery physical examinations report-
                          ing and appearance of clinical reactions

[a]   Dried skim milk.

The remaining subjects of both groups completed the study uneventfully, without any adverse reaction of any kind that could be attributed to the ingestion of the test product. Blood samples were drawn initially and at 60 and 180 d of the study for an extensive battery of biochemical analyses. The experimental group showed no changes or differences of clinical significance relative to those in the control group.

A second study was conducted on the acceptability of and tolerance to Supro® 710. This study was carried out in a larger sample of subjects, as shown in Table 9. All of the subjects completed the study uneventfully and without significant complaints. Gastrointestinal function remained normal throughout the study, and no nausea, abdominal discomfort, loose stools, constipation, or increased gas were reported. No allergic responses were observed, nor were there any changes in blood chemistries in the experimental group compared with those of the control group.

In one of our long-term metabolic studies[6] that included five atopic subjects, we did not observe any changes in the specific immunological response to soy antigens except for a measurable, but slight, increase in soy protein-specific IgG in one nonatopic subject. Goulding et al.[39] reported that 9% of a group receiving at least four main meals per week based on textured soybean protein showed an increase in soy-specific antibody levels. Also, in this long-term metabolic study,[6] there were no significant differences between two groups of subjects, each receiving a different concentrate, in their antisoy antibody titre after the termination of the study. Furthermore, none of the subjects showed any signs, either from laboratory tests or clinical observations, that were indicative of a soy allergy. There was no increase in soy-specific IgE, in contrast to Goodwin,[40] who did report a significant increase in some subjects consuming soy-containing diets during a 4-week period. Additionally, Haeney et al.[41] found that healthy persons had only low activities of antibody to soy, in amounts that reflect presumably no more than harmless exposure. Our collective results confirm these observations.

Most reports of nutritional studies with soy protein products in humans make no specific mention of the acceptability of or tolerance to the material tested. However, if significant problems of intolerance had occurred it is likely that they would have been discussed. We conclude from our experience and from the available literature, therefore, that well-processed soy products consumed over long periods of time are well tolerated and accepted by human subjects.

IX. SUMMARY AND CONCLUSIONS

We have discussed the nutritional value of processed soy protein (isolated soy proteins and soy protein concentrates) in relation to adult human protein nutrition. Well-processed soy protein isolates and soy protein concentrates can serve as the major, or sole, source of the protein intake, and their protein value is essentially equivalent to that of foods of animal origin. These products are fully capable of meeting the long-term essential amino acid and protein needs of adults as well as children (see Chapter 10). The significance of the sulfur amino acid content of soy protein for practical human nutrition is also examined. Under conditions of normal usage of soy protein, methionine supplementation is not only unnecessary, but may even be undesirable. Perhaps for the feeding of the newborn, modest supplementation of soy-based formulas with methionine may be beneficial. At total protein intakes that approximate current dietary protein allowances for the adult, well-processed soy protein products can replace meat and fish proteins without reducing the overall efficiency of utilization of dietary nitrogen. Finally, they are well tolerated and of good acceptability. Thus, soy-based protein foods offer a valuable and nutritious means of achieving a healthful diet.

REFERENCES

1. Energy and Protein Requirements, FAO/WHO/UNU, Tech. Rep. Ser. No. 724, World Health Organization, Geneva, 1985.
2. **Young, V. R. and Pellett, P. L.,** How to evaluate dietary protein, in *Milk Proteins: Nutritional, Clinical, Functional and Technological Aspects,* Barth, C. A. and Schlimme, E., Eds., Steinkopff Verlag, Darmstadt, 1988, 7.
3. Protein Quality Evaluation, FAO/WHO, Report of a Joint FAO/WHO Expert Consultation, Food and Agriculture Organization, Rome, 1990.
4. **Young, V. R., Puig, M., Queiroz, E., Scrimshaw, N. S., and Rand, W. M.,** Evaluation of the protein quality of an isolated soy protein in young men: relative nitrogen requirements and effect of methionine supplementation, *Am. J. Clin. Nutr.,* 39, 16, 1984.
5. **Istfan, N., Murray, E., Janghorbani, M., and Young, V. R.,** An evaluation of the nutritional value of a soy protein concentrate in young adult men using short-term-N-balance method, *J. Nutr.,* 113, 2516, 1983.
6. **Beer, W. H., Murray, E., Oh, S. H., Pedersen, H. E., Wolfe, R. R., and Young, V. R.,** A long-term metabolic study to assess the nutritional value of, and immunological tolerance to, two soy protein concentrates in adult humans, *Am. J. Clin. Nutr.,* 50, 997, 1989.
7. Diet and Health, Food and Nutrition Board, National Research Council, National Academy Press, Washington, D.C., 1989.
8. **Young, V. R. and Scrimshaw, N. S.,** Nutritional evaluation of protein and protein requirements, in *Protein Resources and Technology,* Milner, M., Scrimshaw, N. S., and Wang, D. I. C., Eds., AVI Publishing, Westport, CT, 1978, 136.
9. **Young, V. R., Rand, W. M., and Scrimshaw, N. S.,** Measuring protein quality in humans: a review and proposed method, *Cereal Chem.,* 54, 929, 1977.
10. **Pellett, P. L. and Young, V. R.,** *Nutritional Evaluation of Protein Foods,* Food and Nutrition Bulletin Supplement No. 4, United Nations University, Tokyo, 1980, 154.
11. **Young, V. R.,** Recent advances in evaluation of protein quality in adult humans, in *Proceedings 9th International Congress Nutrition,* Vol. 3, Chavez, A., Bourges, H., and Basta, S., Eds., S. Karger, Basel, 1975, 348.
12. **Hegsted, D. M.,** *Improvement of Protein Nutriture,* Committee on Amino Acids, National Research Council, National Academy of Sciences, Washington, D.C., 1974.
13. **Young, V. R., Scrimshaw, N. S., Torun, B., and Viteri, R.,** Soybean protein in human nutrition: an overview, *J. Am. Oil Chem. Soc.,* 58, 110, 1979.
14. **Torun, B., Viteri, F. E., and Young, V. R.,** Nutritional role of soya protein for humans, *J. Am. Oil Chem. Soc.,* 58, 400, 1981.
15. **Young, V. R.,** Soy protein in relation to adult human protein and amino acid nutrition, in *Soy Protein in Human Nutrition,* Inoue, G., Ed., Research Committee of Soy Protein Nutrition (Japan), Fuji Oil Company, Osaka, 1988, 3.
16. **Wang, M. F., Kishi, K., Takahoshi, T., Komatsu, T., Ohnaka, M., and Inoue, G.,** Efficiency of soy protein isolate in Japanese young men, *J. Nutr. Sci. Vitaminol.,* 29, 201, 1983.
17. **Kaneko, K., Inayama, T., and Koike, G.,** Utilization of soy protein isolate mixed with rice protein in Japanese women, *J. Nutr. Sci. Vitaminol.,* 31, 99, 1985.

18. **Scrimshaw, N. S., Wayler, A. H., Murray, E., Steinke, F. H., Rand, W. M., and Young, V. R.,** Nitrogen balance response in young men given one of two isolated soy protein or milk proteins, *J. Nutr.,* 113, 2492, 1983.

19. **Young, V. R., Wayler, A., Garza, C., Steinke, D. H., Murray, E., Rand, W. M., and Scrimshaw, N. S.,** A long-term metabolic balance study to assess the nutritional quality of an isolated soy protein and beef proteins, *Am. J. Clin. Nutr.,* 30, 8, 1984.

20. **Istfan, N., Murray, E., Janghorbani, M., Evans, W. J., and Young, V. R.,** The nutritional value of a soy protein concentrate (STAPRO-3200) for long-term protein nutritional maintenance in young men, *J. Nutr.,* 113, 2524, 1973.

21. **Wayler, A., Queiroz, E., Scrimshaw, N. S., Steinke, F. H., Rand, W. M., and Young, V. R.,** Nitrogen balance studies in young men to assess the protein quality of an isolated soy protein in relation to meat proteins, *J. Nutr.,* 113, 2485, 1983.

22. **Kies, C. and Fox, H. M.,** Comparison of protein nutritional value of TVP, methionine enriched TVP and beef at two levels of intake for human adults, *J. Food Sci.,* 36, 841, 1971.

23. **Zuzulka, A. and Calloway, D. H.,** Nitrogen retention in men fed varying levels of amino acids from soy protein with and without added L-methionine, *J. Nutr.,* 106, 212, 1976.

24. **Fomon, S. J., Ziegler, E. E., Filer, L. J., Nelson, S. E., and Edwards, B. B.,** Methionine fortification of a soy protein formula fed to infants, *Am. J. Clin. Nutr.,* 32, 21460, 1979.

25. **Fomon, S. J., Ziegler, E., Nelson, S. E., and Edwards, B. B.,** Requirement of sulfur-containing amino acids in infancy, *J. Nutr.,* 116, 1405, 1986.

26. **Ausman, L. M., Gallina, D. L., Hayes, K. C., and Hegsted, D. M.,** Comparative assessment of soy and milk protein quality in infant Cebus monkeys, *Am. J. Clin. Nutr.,* 43, 112, 1986.

27. Energy and Protein Requirements, FAO/WHO, FAO Nutrition Meetings, Dept. Ser. No. 52, Food and Agriculture Organization, Rome, 1973.

28. **Bressani, R., Scrimshaw, N. S., Behar, M., and Viteri, F.,** Supplementation of cereal proteins with amino acids. II. Effect of amino acid supplementation of corn maize at intermediate levels of protein intake on the nitrogen retention of young children, *J. Nutr.,* 66, 501, 1958.

29. **Bressani, R., Wilson, D. L., Behar, M., and Scrimshaw, N. S.,** Supplementation of cereal proteins with amino acids. III. Effect of amino acid supplementation of wheat flour as measured by nitrogen retention of young children, *J. Nutr.,* 70, 176, 1960.

30. **Scrimshaw, N. S., Bressani, R., Behar, M., and Viteri, F.,** Supplementation of cereal proteins with amino acids. I. Effect of amino acid supplementation of corn maize at high levels of protein intake on the nitrogen retention of young children, *J. Nutr.,* 66, 485, 1958.

31. **Kaneko, K., Nishida, K., Yatsuda, J., and Koike, F.,** Effect of methionine supplementation of a soy protein isolate on short-term nitrogen balance in young women, *J. Nutr. Sci. Vitaminol.,* 32, 123, 1986.

32. **Kishi, K., Maekowa, M., Yamamoto, S., Shiquka, F., and Inoue, G.,** Effect of methionine supplementation to soy protein isolate on short-term nitrogen balance in adult men, *Nutr. Sci. Soy Protein (Japan),* 5, 88, 1984.

33. **Koishi, H., Okuda, T., and Miyoshi, H.,** Nutritive effects of L-methionine supplement to the soy protein isolate in human body, *Nutr. Sci. Soy Protein (Japan),* 5, 99, 1984.

34. **Takahashi, R. and Yamada, T.,** Effect of supplementing methionine to soy protein isolate on the protein utilization in male adult humans, *Nutr. Sci. Soy Protein (Japan),* 5, 94, 1984.

35. **Young, V. R. and Pellett, P. L.,** Protein evaluation, amino acid scoring and FDA's proposed food labeling regulations, *J. Nutr.,* 121, 145, 1991.

36. **Taudorf, E., Bundgaard, A., Hancke, S., Hansen, L. V., Prahl, P., and Weeke, B.,** Non-hydrolyzed and hydrolyzed soy protein, *Allergy,* 39, 203, 1984.

37. **Sissons, J. W., Nyrup, A., Kilshaw, P. J., and Smith, R. H.,** Ethanol denaturation of soybean protein antigens, *J. Sci. Food Agric.,* 33, 706, 1982.

38. **Heppell, I. M. J., Sissons, J. W., and Pedersen, H. E.,** A comparison of the antigenicity of soya-bean infant formulas, *Br. J. Nutr.,* 58, 393, 1987.

39. **Goulding, N. J., Gibney, M. J., Gallagher, P. H., Morgan, J. B., Jones, D. B., and Taylor, T. G.,** The immunological consequences of a high intake of soya-bean protein in man, *Qual. Plant. — Plant Foods Hum. Nutr.,* 32, 19, 1983.

40. **Goodwin, B. F. J.,** IgE antibody levels of ingested soya protein determined in a normal adult population, *Clin. Allergy,* 12, 55, 1982.

41. **Haeney, M. R., Goodwin, B. J. F., Barrett, M. E. J., Mike, N., and Asquith, P.,** Soya protein antibodies in man: their occurrence and possible relevance in coeliac disease, *J. Clin. Pathol.,* 35, 319, 1982.

42. **Scrimshaw, N. S., Murray, E., and Young, V. R.,** unpublished data.

Chapter 13

HEART DISEASE — CURRENT STATUS AND DIETARY RECOMMENDATIONS FOR PREVENTION AND TREATMENT

Scott M. Grundy

Heart disease is the major cause of death in most developed countries of the world.[1] Coronary heart disease is responsible for more than 550,000 deaths in the U.S. annually.[2] The incidence of heart disease in the U.S.S.R and European countries has been reported to be at an even higher rate than those in the U.S.[3] It has been estimated by Chazov[4] that 1.5 million individuals in the U.S.S.R. die annually from heart disease. Since heart disease is by far the largest single cause of death in most developed countries, a change in its incidence among the population will significantly influence the health of the overall population. This affects not only mortality, but also morbidity in the population. It is estimated that in the U.S. alone, coronary heart disease is costing 78 billion dollars a year in direct or indirect cost based on the recent National Institute of Health Consensus Development Conference.[2,5,6]

Coronary heart disease is caused by atherosclerosis, which is a chronic disease resulting from progressive changes in the arteries from the accumulation of plaque as a result of depositions of cholesterol and scar tissue in these blood vessels. It has long been known that the incidence of heart disease is much lower in some populations and countries than in others.[7] Epidemiological studies of the movement of populations from one country to another, however, identify that the differences are due to cultural effects rather than population genetic effects.[7,8]

Epidemiological studies have also identified a number of factors which relate to the occurrence of heart disease.[2] Individual family genetics plays an important factor in the development of the disease. Other important factors which have been identified include the occurrence of high serum cholesterol levels, smoking, high blood pressure, obesity, and diabetes.

The occurrence of elevated serum cholesterol has been associated with an increased risk of coronary heart disease. A comparison of the serum cholesterol levels of populations in southern Japan and eastern Finland is illustrated in Figure 1. The cholesterol levels are quite different between these two populations. The cause of this difference of course could be multifactorial, but certainly the differences in diet consumption are a major factor. The incidence of coronary heart disease is low in southern Japan and very high in eastern Finland. The dietary patterns in these two areas are also quite different, with the Japanese consuming a high vegetable/fish diet and the Finns consuming a high-fat, animal-products diet.

The relationship of serum cholesterol levels and 6-year death rates per 1000 men in the 35 to 57 age range is summarized in Figure 2 from the report of Martin et al.[9] There is a direct relationship between the level of cholesterol in the blood and the incidence of death from heart disease among men. In particular, the death rate increases significantly above 200 mg/dl and the incidence increases over the range of cholesterol levels up to 300. These data clearly suggest that if total cholesterol levels and low-density lipoprotein (LDL) cholesterol are reduced, there will be a potential for reducing the incidence of death from heart disease by a factor of up to four times.

Table 1 compares the risk of coronary heart disease death per 1000 men 35 to 57 years of age by quintiles, starting with 182 mg/dl or less and going up to greater than 245 mg/dl. The increase in the death rate over this range is progressive for the nonsmoking normotensive individual. The increase in the death rate between the lowest quintile and the highest quintile of the normotensive individuals is a factor of four.

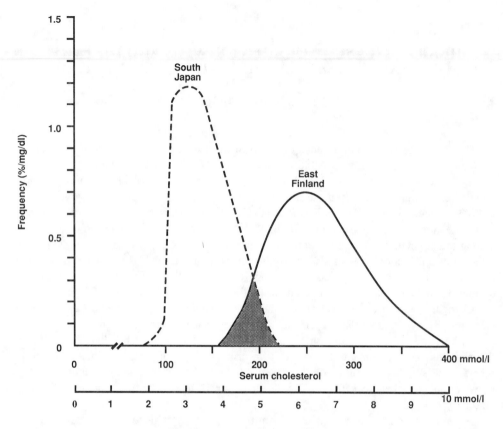

FIGURE 1.   Cultural differences in serum cholesterol levels.

If the same comparison is made with the hypertensive smoker, the effects are similar between the lowest quintile and the highest quintile. In addition, there is an increase by approximately a factor of four between the normotensive nonsmoker and the hypertensive smoker at the same quintile. Both the level of serum cholesterol and the other environmental factors, such as smoking and hypertension, can magnify the incidence of heart disease. Therefore, a multifactional approach to the prevention of coronary heart disease will optimize the prevention of this disease in the population.

A number of factors can interplay on the development of heart disease, which have as their origin a dietary component. These include obesity, which further adds to the risk of heart disease and diabetes and has been strongly linked to an increased incidence of heart disease in individuals having this problem. An important recommendation that has been made by the National Institute of Health Consensus Conference[2] is to optimize normal weight through proper dietary managment to avoid excess consumption of calories.

The primary dietary factors that have been identified and the modifications that were recommended are related to fat intake and the types of fats consumed. As early as 1957, the American Heart Association Nutrition Committee[10] recommended dietary modifications to reduce fat intake and the intake of saturated fat, and these recommendations have been further defined in subsequent reports.[11-16] Over the 30-year period since the introduction of these recommendations, there has been a 25% reduction in heart-related deaths in the U.S. Some of this is due to improved medical care, but a large portion is due to a change in risk factors, which include a reduction in smoking, diet modification, and exercise.

The present intake of fat in the U.S. diet, as well as in other developed countries, exceeds 40% of calories. A reduction in fat intake from 40 to 30%, if it is accompanied by

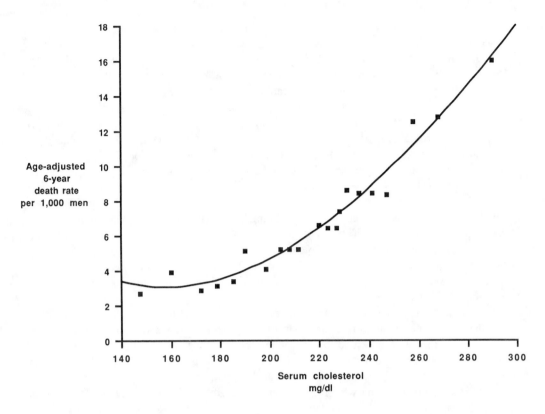

FIGURE 2.   Relationship of serum cholesterol to coronary heart disease in 361,662 men age 35 to 67 during an average follow-up of 6 years.

TABLE 1
Coronary Heart Disease Deaths Per 1000 in Men Age 35 to 57 with an Average Follow-up of 6 Years According to Serum Cholesterol Quintile and Presence or Absence of Other Risk Factors

Serum cholesterol quintile (mg/dl)	Normotensive nonsmoker	Hypertensive smoker
<182	1.6	6.3
182—202	2.5	10.0
203—220	2.7	15.5
221—244	3.8	16.6
≥245	6.4	21.4

Note: The difference in absolute risk in the highest vs. the lowest quintile of serum cholesterol is greater in patients who are at high risk for other reasons.

a decrease in saturated fatty acids, will reduce serum cholesterol levels. Originally it was felt that a shift in fat type from saturated to polyunsaturated would further benefit the population. It now appears that the primary benefits in cholesterol lowering are in reducing the amount of saturated fat to less than 10% of total calories. Recent studies[18] in our laboratory suggest that monounsaturated and polyunsaturated fats may have essentially equivalent cholesterol-lowering ability in replacing saturated fat. Other recommendations include reducing the cholesterol intake in the diet to less than 300 mg/d.

In individuals with very high cholesterol levels (in excess of 240 mg/dl, it is recommended that they go to a step-two diet which will reduce the saturated fats to less than 7% of calories and the cholesterol levels to less than 200 mg/day.

The National Institutes of Health has recently initiated a National Cholesterol Education Program for the reduction of serum cholesterol levels.[19] This program is designed to implement the detection, evaluation, and treatment of high cholesterol levels in individuals and to reduce average cholesterol levels in the general public. This program will take two directions. One program is to inform and educate physicians to aggressively detect high serum cholesterol levels and advise their patients on dietary modifications based on analysis of blood cholesterol. The second program is to educate the public on the importance of maintaining low serum cholesterol levels and the appropriate dietary programs to reduce serum cholesterol levels. To accomplish this, the program is emphasizing that everyone have a blood cholesterol measurement. Testing in the physician's office is preferred, but rapid, simple cholesterol analysis methods are now being taken to the general population in shopping centers to facilitate the identification of individuals with elevated cholesterol levels who are at greater risk. These latter measurements, however, require interpretation by physicians. The primary treatment for high serum cholesterol levels is dietary modification. Failure to respond adequately to dietary modification in individuals with very high cholesterol levels may make drug treatment necessary.

Since both dietary modification and drug treatment are lifetime programs, it is important that the dietary modification become a part of the general understanding of the population. The consumer must be educated to identify the food products that will provide the type of diet needed to reduce serum cholesterol. These foods must also be available for the consumer to purchase as well as be clearly identified for their purpose and benefit.

Basic recommendations which the consumer can follow include reduction in high-fat food products. These include reductions in high-fat animal products; reduction in high-saturated fat dairy products, and whole eggs and organ meats which contain considerable amounts of cholesterol, is also required. In addition, it is important that food products be cooked with a minimum of fat added to the diet. These programs are long-term projects that require an ongoing education process for the consumer, the food processor, and the physician to further reduce the incidence of heart disease in the population.

REFERENCES

1. Prevention of Coronary Heart Disease, Report of a WHO Expert Committee, Tech. Rep. Ser. No. 678, World Health Organization, Geneva, 1982.
2. Lowering blood cholesterol to prevent heart disease, National Institute of Health Consensus Conference, *J. Am. Med. Assoc.*, 253, 2080, 1985.
3. World Health Organization, Regional Office for Europe, Myocardial Infarction Community Register (Public Health in Europe, No. 5), Copenhagen, 1976.
4. **Chazov, E. I.,** Problems of prevention from the position of specialization and integration, *Ter. Arkh.*, 55, 5, 1983.
5. *Heart Facts,* American Heart Association, Dallas, 1986.
6. Cost of Illness and Disease — Fiscal Year 1975, Report #B1, Public Services Laboratory, Georgetown University, Washington, D.C., 1977.
7. **Syme, S. L., Marmot, M. G., Kagan, A., Koto, H., and Rhoads, G.,** Epidemiological studies of coronary heart disease and stroke in Japanese men living in Japan, Hawaii and California, *Am. J. Epidemiol.*, 102, 514, 1975.
8. **Koto, H., Tillatson, J., Nickamon, M., Rhoads, G., and Hamilton, H. B.,** Epidemiological studies of coronary heart disease and stroke in Japanese men living in Japan, Hawaii and California, *Am. J. Epidemiol.*, 97, 372, 1973.
9. **Martin, M. J., Hulley, S. B., Browner, W. S., Kuller, L. H., and Wentworth, D.,** Serum cholesterol, blood pressure and mortality: implications from a cohort of 361,682 men, *Lancet*, 2, 933, 1986.

10. **Page, I. H., Stare, F. I., Corcoran, A. C., Pollack, H., and Wilkinson, C. F.,** Atherosclerosis and fat content of the diet, *Circulation,* 16, 163, 1957.

11. Central Committee for Medical and Community Program of the American Heart Association, Dietary fat and its relation to heart attacks and strokes, *Circulation,* 133, 1961.

12. Diet and Heart Disease, American Heart Association, Dallas, 1965.

13. Diet and Heart Disease, American Heart Association, Dallas, 1968.

14. Diet and Heart Disease, American Heart Association, Dallas, 1973.

15. Diet and Heart Disease, American Heart Association, Dallas, 1978.

16. **Grundy, S. M. et al.,** Rationale of the diet-heart statement of the American Heart Association, Report of Nutrition Committee, *Circulation,* 65, 839a, 1982.

17. **Grundy, S. M. et al.,** Coronary Risk Factor Statement for the American Public, Report of Nutrition Committee, American Heart Association, 1985.

18. **Grundy, S. M.,** Comparison of monounsaturated fatty acids and carbohydrates for lowering plasma cholesterol, *N. Engl. J. Med.,* 314, 745, 1986.

19. National Cholesterol Education Program Report of the Expert Panel on Selection, Evaluation and Treatment of High Blood Cholesterol in Adults, National Heart, Lung and Blood Institute, National Institutes of Health, Bethesda MD, 1987.

Chapter 14

CARDIOVASCULAR DISEASE AND OBESITY IN THE U.S.S.R.

Michael A. Samsonov and Valentina A. Meshcheryakova

The significant shifts which took place in many areas of society throughout the 20th century have also led to considerable changes in human nutrition which many researchers[1-3] believe have an impact on the structure of disease. Widespread diseases caused by too much food in economically developed countries is a depressing sign of the times. Statistics show obesity to rank among the highest of the so-called excessive food diseases.

The data reported by McCracen[4] indicate that 20% of all adults in the U.S. are obese. Obesity is encountered in roughly 40% of all adults living in the cities of Czechoslovakia,[3] while it measures an average of 11.6% among children. In addition, a direct correlation was observed between the development of obesity and an elevated caloric content in food due to an increased content of fats and carbohydrates in a serving of food as well as reduced physical activity. In England, 20 to 40% of all adults are reported to be overweight.[5] In France, 50% of the population aged 40 and up are 10% or more overweight. Excessive body weight and obesity are encountered less often in Japan, occurring in 10% of the women and 8% of the men according to data compiled by several authors.[6]

Both excessive body weight and obesity are often accompanied by atherosclerosis, myocardial infarction, hypertention, cholelithiasis, and sugar diabetes, which are frequently the cause of death. Calculations conducted by a U.S. insurance company showed that obese individuals die 7 to 10 years earlier than people with a normal body weight.

Obesity is becoming a serious social problem, since it facilitates the development of serious diseases (e.g., atherosclerosis, ischemic disease, diabetes, etc.) as well as diminishes the fitness of a population for work and creates the preconditions for the development of obesity by the next generation. According to statistics,[7] death due to cardiovascular disorders is 1.5 times more frequent in obese individuals than in people who eat normal or reduced amounts of food.

Data compiled by various authors indicate that damaged blood vessels are encountered in about 10 to 54% of individuals suffering from obesity.[8] Of the accompanying disease observed in obese individuals according to data from the Institute of Nutrition of the Soviet Academy of Medical Sciences, atherosclerosis was diagnosed in 26.5% of the cases and hypertension in 22.5%.[9]

The dependence between the incidence of atherosclerosis and coronary disease and the extent of disruptions to lipid metabolism, such as hypertriglyceridemia and hypercholesterolemia, may be viewed as established based on the sweeping epidemiological examinations carried out in several countries, and in particular on the Framingham experiment. The disruptions in cholesterol metabolism which accompany increases in blood and bile cholesterol content are related to a certain extent to the etiology of cholelithiasis. According to data compiled by insurance companies in the U.S.,[7] two to three times as many individuals with excess weight developed cholelithiasis and died of it than people within normal weight limits. There is no doubt about the correlation between obesity and diabetes. Obesity is accompanied by a reduced tolerance to glucose in 60 to 80% of the obese individuals, and in a considerable number of cases (5 to 20%) by pronounced symptoms of diabetes.[8] The dependence between excess consumption of nucleic acids as well as disruptions to purine metabolism and the risk of contracting diseases such as gout, metabolic polyarthritis, and symptoms of cholelithiasis represents proof of the above observations.

One of the factors at work in the development of obesity is the reduction in energy expended by modern man due to the mechanization and automation of production processes, the expansion of city transportation networks, the consistent rise in automobile production, and the construction of skyscrapers which invariably come complete with elevators. In short, the paradox in modern times lies in the fact that such a positive phenomenon as technological progress has brought with it the negatives of physical stress and reduced physical activity.

Another important factor helping to increase body weight is overeating. Taken together, all of the above factors frequently end up doing severe damage to the health. It is precisely in light of this fact that the development of measures for preventing and treating obesity is currently taking on social and medical significance. Due to the extreme multiplicity of factors and the complexity involved in studying the impact which the kind of diet has on the health of the population, many aspects of this issue remain the topic of persistent study and scientific debate.

Epidemiological examinations play a major role in any studies which are aimed at ascertaining the prevalence of obesity as well as determining those factors which serve to increase weight gain. They are comprehensive and provide invaluable assistance in finding out the effect exerted by the type of diet on the health of a population. Epidemiological studies are especially important due to the fact that accurate statistical data concerning the prevalence of obesity in the U.S.S.R. were absent until recently. It was for this reason that epidemiological studies were carried out in various regions of the U.S.S.R. under the direction of the Institute of Nutrition of the Soviet Academy of Medical Science jointly with several peripheral medical institutes. Broad segments of the population were included in the observation. The average number of individuals examined was in excess of 38,000.

Excess body weight as well as its extreme form — obesity—were found in 50% or more of all adults (especially women). Various levels of obesity here make up nearly 26%, or half of all of the overweight people. It was demonstrated by analyzing the data according to gender that the given pathology prevailed in females, with the pattern persisting throughout the entire group investigated. Obesity was encountered in only 4.5% of male industrial workers, for example, while 21.5% of the women were found to be overweight. The same phenomenon may be observed at enterprises of the food industry, as well, where 6.8% of the men were found to be obese, while this percentage jumped sharply to 29.2% among women. We may assume from the data that from 1.5 to 3 times as many women than men suffered from obesity, which corresponds to data in published literature.

The tendency toward increased obesity among the young was particularly alarming, where 14% were found to be obese. A clearly progressive rate of obesity was found in individuals after the age of 35. The distribution of obesity by profession showed that the disease afflicted a significantly high percentage of individuals across the board. Obesity is most often encountered among workers in the food industry. The least incidence was found at collective farms and in agricultural workers.

The role played by the factors of age, sex, and profession in the advent and progression of obesity is abundantly clear. We were able to conclude from an analysis conducted on the diets of people living in various geographical regions of the U.S.S.R. that energy imbalance was a leading factor in the development of obesity. This term encompasses both high-calorie foods and decreased physical activity, which manifests itself in a lack of equilibrium between the amount of energy consumed and expended. The deviations primarily involve the excess consumption of fats and oligosaccharides.

An analysis of the diets of over 1000 obese persons coupled with data from an epidemiological examination of several thousand individuals in Moscow and outlying areas confirms that the majority of overweight people eat too much. The calorie content in the foods is as high as 5000 to 6000 kcal per day. In 47% of the obese individuals the pattern of food consumption was imbalanced during the day by consuming the majority of calories in the evening hours. The energy overload in the body activates the synthesis of fatty acids, triglycerides, and cholesterol, and the dynamic equilibrium between lipolysis and lipogenesis is shifted in favor of the latter. The ingestion of excessive amounts of food may serve as a specific indicator that certain parts of the metabolism of the body are weak and explain why adaptive mechanisms become drained and persistent disruptions of the metabolic processes set in.[1,9]

An entire set of metabolic disruptions must be taken into account when developing scientifically based recommendations concerning dietary regimes for the prevention and treatment of obesity. One very important and promising area has to do with determining the enzymatic activity of the fatty tissue in obese people. Studies done by several authors[1] demonstrated that the level of lipase and lipoprotein lipase activity in fat individuals is substantially lower than the norm (5 and 2.6 times, respectively). The reduced activity of enzymes was also observed with regard to specific activators (adrenalin, heparin). The fatty tissue, therefore, becomes metabolically inert in the obese.

Within these examinations is the study of another enzyme in the fatty tissue — glucose-6-phosphate-dehydrogenase (G-6-P-DH). A considerable drop in G-6-P-DH activity was discovered in individuals with alimentary obesity. While averaging 20.42 μmol/min per 1 g of protein in healthy persons, G-6-P-DH activity measured 5.5 μmol/min per 1 g of protein following treatment. One of the enzymes which provides the quantity of reduced NADP necessary for the biosynthesis of fatty acids is also diminished. The above facts enable us to imagine a unique kind of maladaptation which disrupts lipid homeostasis as being one of the possible mechanisms behind the development of obesity.[10]

In order to prevent this disease, serious attention must be focused on studying the diets of individuals suffering from obesity at a young age. Even in the first stage of obesity, overweight young people cannot elude the shifts in metabolism which are common for all victims. In a significant number of obese individuals, in addition to the abnormalities in fat and carbohydrate metabolism, were found type II, III, and IV hyperlipoproteinemia (according to Fredrikson[64]). This points to the presence of atherogenic lipoproteins and is pathogenetically connected to the early onset of atherosclerosis in obese people. Only 26.4% of the subjects examined had no hyperlipemia and exhibited a normal lipoprotein spectrum. The remainder revealed various types of hyperlipoproteinemia, with type IIa occurring in 7.6%, type IIb in 27.4%, type IIab in 17.7%, type II in 11%, and type IV in 9.9% of the people.

Therefore, inherent in obesity are changes in the body as a result of disruptions in the regulatory mechanisms and shifts in metabolic processes. The variety of different kinds of obesity coupled with the stubborn persistence of the disease illustrates the necessity for developing pathogenetically substantiated methods of treatment. One major step forward in the area of therapeutic diets can be seen in the newly conceived theory which views a balanced diet as being the number one precondition for developing all types of menus with consideration of the real requirements of the afflicted individual, the details behind the metabolic processes, and the condition of separate function systems. One of the jobs in therapeutic diets in general and in obesity in particular could involve restoring the disrupted qualitative and quantitative balance between the enzymatic systems of the organism and the chemical structures of food by adapting the caloric value and chemical composition of the rations and the foods to the metabolic needs of the organism.[1]

The role played by the dietary factors in the etiology and pathogenesis of the disease remains pivotal regardless of the kind of obesity at issue. This is why the physician must in all cases focus primary attention on the diet consumed by the obese patient. Obesity occurs only when the energy taken in from the food exceeds the amount of energy expended. This fact allows us to formulate dietary principles for treating obesity, the first of which involves prescribing a reduced-calorie regimen. Another important factor to consider is balance within the food products in terms of the actual requirements of the organism for protein, polyunsaturated fatty acids (PUFA), and products high in biological value. The steps envisaged are as follows: (1) limiting insulinogenic substances, in particular simple sugars; (2) including in the diet relatively large amounts of vegetable fats with the objective of activating the lipolytic system; (3) creating a satiated feeling with a low-calorie diet high in bulk foods (fruits and vegetables); (4) eating five to six times a day with the objective of training the enzyme systems and reducing the appetite; (5) using so-called zigzags in the diet (high-contrast, unloading days). Limiting the intake of water and table salt is one of the most important principles. Care must be taken when limiting the amount of water ingested to take into account the possibility that products from incompletely metabolized fats might accumulate in the organism to produce ketogenesis. Prescribing alkaline mineral water — Borzhomi in particular — might be of some importance in preventing these events from occurring as well as maintaining the acid-base equilibrium. The final step has to do with evaluating the effectiveness and correcting the treatment based on a relatively thorough study of the biochemical status of every patient.

The primary complication in the treatment of obese individuals is the long time required to see the results, while immediate results are usually the ones which reassure people to stay on a treatment program. We are talking not just about losing a sufficient amount of weight, but also about the effect on several important parameters of metabolism — positive shifts in the lipid indices for the blood (general lipids, cholesterol, lipoproteins, NEFA) as well as in the fatty acid composition of the serum. Specifically, the level of eicosatrienic acid (C 20:3) elevated nearly by a factor of 2 goes down, the percentage of atherogenic IV and lipoprotein fractions drops, the fat-mobilizing lipase of the fatty tissue becomes activated, and the volume of fat cells falls. Several indices for carbohydrate metabolism — the glycemic and unesterified (free) fatty acids curves and water metabolism — improve at the same time.[9]

Those suffering from obesity at first tend not to visit their physician, since their overall condition remains satisfactory for a long period of time, even though significant deviations in the metabolism of substances take place in this period as well, primarily in lipid metabolism. What the physician most often ends up seeing are obese individuals in a period when the primary complaints are dyspnea, cardialgia, elevated arterial pressure, and the appearance of edema. At this point, diseases are present which frequently result from obesity, and for that reason determine the prognosis for the patient in terms of his or her job and life in general.

As indicated above, the diseases which most frequently accompany obesity are ischemic heart disease (IHD), myocardial infarction, and hypertension. The most common pattern pathogenetically linking obesity with these diseases is disrupted lipid metabolism, whose restoration through body weight reduction represents the essence of the therapeutic treatment. Clinical experience shows us that comprehensive methods for treating patients with IHD and hypertension under conditions of both excess and normal weight have been developed through clinical experience over the last 25 years in developed countries of the world, and in particular the U.S.S.R.

A therapeutic diet based on metabolic principles represents one of the primary factors in the comprehensive treatment of these patients. The experience gained by Soviet scientists[11] has made it possible to develop the basic principles for the dietetic treatment of patients with IHD and hypertension. These principles are based on:

1. Diets in which the caloric value corresponds to the amount of energy expended
2. Reducing the amount of fat in the diet, particularly saturated fats
3. Reducing mono- and disaccharides
4. Reducing sodium ions (sodium chloride)
5. Reducing products which contain cholesterol

6. Reducing purine substances
7. Supplementing the diet with PUFA, methionine, choline, and fiber
8. Completely satisfying the requirements of the organism of the patient with vitamins, mineral substances, and microelements
9. Systematic therapeutic exercise, light sports activity such as measured walking

Particular attention was paid during the development of these principles to reducing body weight and restoring the disrupted lipid metabolism. This is why the optimal quantitative and qualitative proportions of fats and carbohydrates in the diet were determined first. At the same time, it is obvious that too little attention was given to the quantitative and particularly the qualitative composition of protein in the dietary ration.

An experiment on rabbits revealed the atherogenic influence exerted by animal proteins — meat, eggs, milk — back at the turn of the century already.[12] There is at present an abundance of convincing epidemiological, experimental, and clinical observations attesting to the close correlation between the quantity and quality of protein in a ration of food and the death rate from IHD,[13-20] as well as the level of cholesterolemia,[21,22] and the extent of atherosclerotic changes in the aorta.[15,23-26]

The source of food protein has a considerable impact on the blood cholesterol content. The most pronounced cholesterolemia is observed when using skim milk, lactalbumin, or casein as the source of protein, while it is less pronounced when employing meat (beef, pork) and negligible when utilizing egg white and wheat gluten. By contrast, the blood cholesterol content was found to decrease in those cases where isolated soy protein served as the source of protein.[21] Many authors have noted the hypocholesterolemic effect exerted by soya protein in both experimental and clinical observations.[14,15,25-32]

Several researchers failed to turn up any reduction in the blood cholesterol level when incorporating isolated soya protein in the diet.[28,33,34] The hypocholesterolemic effect of soya depends to a large extent on other diet-related factors, such as the amount of fats and the exogenic cholesterol content, the qualitative composition of fibers. The chemical structure may either potentiate or level off the hypocholesterolemic effect exerted by the soya protein. In addition, the amino acid imbalance of the ration — a high content of lysine, a shortage of essential amino acids (tryptophan in particular), and a change in the ratio of arginine and lysine — has an effect on the level of lipemia and serves as an explanation for the difference in the effect exerted by casein and soya protein on the blood cholesterol level.[35-38] Several researchers attribute the hypocholesterolemic effect of vegetable proteins to the influence of the saponins contained therein.[39-41] Regardless of the mechanism, the hypocholesterolemic effect of soya protein is determined in large part by the quantity of the latter, as well as by the content and ratio of other nutrients in the ration.

The objective of our examinations was to run a comparison study on two isocaloric variants of antiatherosclerotic diets containing nearly 120 g of protein. Employed in one of these diets (AB) were traditional sources of protein with an animal protein to vegetable protein ration of 2:1. In the second diet (ABC), nearly 15% by weight of the protein was replaced by an isolated soya protein (20 g per day), including in baked bread. The animal to vegetable protein ratio measured 1:1 in the ABC diet. The amino acid composition of the rations employed is shown in Table 1. The main difference in the amino acid composition of the rations are in the content of lysine, which was noticeably higher in the AB diet. Differences in the content of other amino acids were less marked. The content and qualitative composition of fats and carbohydrates in AB and ABC rations were identical.

Observations were carried out on 109 male patients with chronic IHD (CIHD), ranging in age from 36 to 60, who had suffered previous myocardial infarctions. Of these, 71 patients underwent treatment with the ABC diet, while the remaining 38 were put on the AB diet. The dietary treatment lasted for 4 weeks, with drug treatments over this period restricted to

TABLE 1
Comparative Amino Acid Compositions of Antiatherosclerotic Diet

	Traditional diet AB		Antiatherosclerotic diet ABC	
Amino acid	mg/d	mg/g total A.A.	mg/d	mg/g total A.A.
Essential				
Histidine	3,480	29.2	3,176	27.0
Isoleucine	5,554	46.6	5,401	45.9
Leucine	9,690	81.2	9,145	77.7
Lysine	11,522	96.6	7,830	66.5
Methionine	2,879	24.1	2,461	20.9
Cystine	1,733	14.5	1,852	15.7
Methionine + cystine	4,612	38.7	4,313	36.6
Phenylalanine	5,463	45.8	5,501	46.7
Tyrosine	4,477	37.5	4,502	38.2
Phenylalanine + tyrosine	9,940	83.3	10,003	85.0
Threonine	5,020	42.1	4,693	39.9
Tryptophan	1,468	12.3	1,462	12.4
Valine	6,354	53.3	6,086	51.7
Total essential	51,482	431.7	49,209	418.0
Nonessential				
Alanine	5,966	50.0	6,738	57.2
Arginine	6,502	54.5	6,738	57.2
Aspartic acid	10,036	84.1	10,232	86.9
Glycine	5,168	43.3	4,855	41.2
Glutamic acid	22,340	187.3	24,649	209.4
Proline	7,948	66.6	7,617	64.7
Serine	5,712	47.9	5,781	49.1
Total nonessential	67,781	568.3	68,516	582.0
Total all amino acids	119,263	1,000.0	117,725	1,000.0

prescribed coronarolytic agents when indicated by clinical observations. Both diets were well tolerated by the patients, with no major shifts in body weight being observed during the course of the dietary treatment.

Blood samples were taken on two occasions, before and after the diet treatment following 14 h without food intake. The level of cholesterol,[42] triglyceride,[43] and fibrinogen[44] in the blood serum was determined, as were the time of euglobulin lysis,[45] lysis zones using standard fibrin plates,[46] the content of antiplasmins,[47] the partial thromboplastin time, the capillary permeability to water and protein,[12] the activity of prekallikrein and kallikrein,[48] alpha$_2$-macroglobulin,[49] the enzyme activity in the pancreas,[50,51] and the immunological reactivity.[19,20] The Student t test was employed for statistically processing the data obtained.

Over the course of diet therapy, a tendency was observed for the blood cholesterol and triglyceride levels to drop in both groups of patients. Adding soya protein to the ration significantly strengthened its hypocholesterolemic action. Differences in the effect exerted by the two ration diets were observed to vary with the types of hyperlipoproteins (HLP) when comparing the changes in the blood lipid indices. In Table 2 are summarized changes for the blood lipid indices of the patients examined with different types of HLP during diet treatment with diets AB and ABC.

A significant drop in the level of blood cholesterol was observed in types IIa and IIb of HLP. In the type IV HLP, both diets with elevated protein content facilitated an increase in the blood cholesterol level, which was more pronounced in the AB diet. The level of triglyceride in the blood of CIHD patients with type IV HLP fell sharply in the case of the AB ration, while this occurred in patients with type IIb HLP in the ABC ration. In addition to this, a marked reduction was observed in the level of low-density lipoproteins (27% of

TABLE 2
Changes in Blood Lipid Indices in IHD Patients with Various
Types of HLP Before (I) and After (II) Diet Therapy

Parameters (mm/l)	Standard diet AB	Modified diet with soy protein ABC
Type IIa hyperlipoproteinemia		
Cholesterol		
Before	9.4 ± 0.22	8.3 ± 0.25
After	9.8 ± 0.61	7.3 ± 0.28
Change (%)	+4.5	−11.0[a]
Triglycerides		
Before	1.18 ± 0.23	1.12 ± 0.10
After	1.37 ± 0.26	1.13 ± 0.18
Change (%)	+16	+0.9
Type IIb hyperlipoproteinemia		
Cholesterol		
Before	7.77 ± 0.21	8.81 ± 0.4
After	7.39 ± 0.29	7.35 ± 0.34
Change (%)	−4.9	−16.6[a]
Triglycerides		
Before	3.17 ± 0.30	3.03 ± 0.27
After	2.12 ± 0.47	1.40 ± 0.16
Change (%)	−33	−54[a]
Type IV hyperlipoproteinemia		
Cholesterol		
Before	5.3 ± 0.18	5.73 ± 0.11
After	6.8 ± 0.42	6.66 ± 0.23
Change (%)	+28[a]	+16[a]
Triglycerides		
Before	3.30 ± 0.36	2.0 ± 0.08
After	1.58 ± 0.36	1.84 ± 0.14
Change (%)	−52[a]	−8

[a] Statistically significant ($p < 0.05$).

the initial level) and very low-density lipoproteins (13% of the initial level) in patients receiving the ABC diets.

As demonstrated by our previous examinations, increasing the proportion of protein in the ration to 120 g per day (AB variant) noticeably reduces the therapeutic effect exerted by the antiatherosclerotic diet on the blood hypercoagulation syndrome in CIHD patients.[2] The data obtained from these studies (Table 3) indicate that the ABC diet has a more positive effect on the antiatherosclerotic diet on the hemocoagulation system, the lysis time for the euglobulin clot, the lysis of fibrin plates, the partial thromboplastin time, and the content of antiplasmins than does variant AB. The tendency of the level of fibrinogen in the blood serum to rise is maintained in the case of ration ABC as well. Hemocoagulation activity is closely tied to the activity of the kallikrein-kinin system of the organism.[52-54] A significant (more than threefold) increase in the kallikrein level and a reduction (threefold) in the level of alpha$_2$-macroglobulin inhibitor in the blood serum in comparison with healthy individuals was observed in the CIHD patients. As with the trypsin/trypsin inhibitor ration, the tryptic activity displayed by the pancreas did not differ substantially from the level in healthy people. The content of prekallikrein was slightly lower (by 17%) in addition to the level of antitrypsin (1.5 times) in comparison to that common in healthy individuals, while the esterase activity (ethyl benzoylarginine) was somewhat elevated (by 17%). The results obtained might have resulted from the fact that more kinin was formed than decomposed during IHD.

TABLE 3
Changes in the Indices for the Coagulating and Fibrinolytic Systems in IHD Patients Before and After Diet Therapy

Parameters	Standard diet AB	Modified diet with soy protein ABC
Fibrinogen (mg%)		
Before	531 ± 34.7	436 ± 23.4
After	543 ± 41.3	451 ± 24.0
Change (%)	+2	+3
Hydrolysis time, euglobulin clots		
Before	146 ± 7	124 ± 7.0
After	125 ± 7	122 ± 4.2
Change (%)	−14	−2
Hydrolysis of fibrin plates (mm²)		
Before	26 ± 3.0	28 ± 3.5
After	23 ± 3.2	32 ± 2.9
Change (%)	−12	+14
Plasminogen activator (mm²)		
Before	13.0 ± 4.1	25 ± 2.9
After	14.5 ± 3.5	25 ± 3.7
Change (%)	+11	+13
Partial thromboplastin time		
Before	75.0 ± 7.3	77 ± 3.2
After	74.3 ± 6.8	83 ± 4.6
Change (%)	−1	+8
Antiplasmins (%)		
Before	166.5 ± 18.4	214 ± 34.0
After	181.5 ± 21.0	217 ± 25.8
Change (%)	+9	+1

In addition to the tendency to increase the fibrinolytic activity of the blood with the ABC diet, a significant drop in the level of kallikrein (from 139.6 ± 7.9 to 62.2 ± 6.8 meq/ml) and an increase in the content of alpha$_1$-antitrypsin (from 43.6 ± 1.6 to 49.9 ± 2.4 meq/ml; $p < 0.05$) in the blood serum was observed. Vasoactive peptides have a considerable effect on the permeability of vessel walls and the development of inflammatory and degenerative changes. Drugs which exert an antibradykinic effect (anginin) not only retard the development of experimental atherosclerosis, but improve the clinical course taken by a disease as well. In our observations, the increase in the activity of the kallikrein-kinin system in CIHD patients was accompanied by a considerable elevation in the permeability of the capillaries, up from a normal level of 4.8% to $12.3 \pm 1.8\%$ for protein and from a normal 4 ml to 13.5 ± 0.4 ml in the case of water.

The permeability of capillaries to water fell to 11.7 ± 0.5 ($p < 0.05$), along with the drop in the activity of the kallikrein-kinin system during treatment with the ABC diet. There were no significant changes in the permeability of the capillaries with respect to protein. Presented in Table 4 are the indices for immunological reactivity in the patients examined during diet therapy with the ABC diet. The marked reduction in the indices in question points to a drop in the autoimmune shifts which accompany IHD.

Replacing nearly 15% of the overall quantity of protein with isolated soya protein in an antiatherosclerotic diet therefore increases the hypolipemic effect of the diet considerably, primarily by reducing the level of blood cholesterol in CIHD patients with type II HLP. The mechanism involved in the hypocholesterolemic effect exerted by soya protein may vary. Data have been obtained which indicate that soya protein increases the excretion of

TABLE 4

Change in the Indices for Immunological Reactivity in IHD Patients During Therapy with the Isolated Soy Protein-Containing Diet (ABC)

Parameter	Before treatment	After treatment	Control (normal level)
Lymphocyte blastotransformation reaction (%)			
PHA	50 ± 4.0	55 ± 5.0	60 ± 6.0
KonA	54 ± 6.0	50 ± 4.0	67 ± 6.0
HTE	9 ± 0.5	5 ± 0.3[a]	6 ± 0.2
AAE	9 ± 0.4	6 ± 0.02[b]	5.55 ± 0.4
Index of inhibition of Leukocyte migration during exposure to			
PHA	0.80 ± 0.03	0.85 ± 0.02	0.85 ± 0.03
HTE	0.78 ± 0.03	0.84 ± 0.02[b]	0.85 ± 0.03
AAE	0.76 ± 0.04	0.82 ± 0.03[b]	0.85 ± 0.03
Index for cytotoxicity of lymphocytes			
PHA	0.42 ± 0.01	0.36 ± 0.02	0.35 ± 0.01
HTE	0.23 ± 0.01	0.18 ± 0.01[b]	0.15 ± 0.01
AAE	0.26 ± 0.02	0.19 ± 0.02[b]	0.15 ± 0.02
Concentration of immunoglobulins (Ig) in blood serum			
Ig G (mg%)	1496 ± 31.0	1372 ± 27.0[b]	1268 ± 30.0
Ig M (mg%)	160 ± 3.0	149 ± 4.0	143 ± 10.0
Ig A (mg%)	210 ± 9.0	183 ± 10.0	225 ± 14.0
Ig D (mg%)	4.6 ± 0.2	3.5 ± 0.1[b]	3.3 ± 0.2
Ig E (μ/ml)	59 ± 6.0	55 ± 7.0	54 ± 5.0

Note: HTE, water-salt heart tissue extract; AAE, water-salt extract from an atherosclerotically altered aorta; PHA, phytohaemagglutinin; KonA, concanavalin A.

[a] $p < 0.001$.
[b] $p < 0.05$.

sterol with the feces and reduces the intestinal absorption and reabsorption of cholesterol.[37-41] Published information is available about soya proteins accelerating cholesterol metabolism and increasing the activity of 3-hydroxy-3-methylglutaryl-CoA-reductase in the liver.[22,55,56] Reducing the content of lysine in a ration of ABC may have an effect on the synthesis of various classes of apoproteins, which explains the differences in the effect exerted by soya protein on the blood cholesterol level in different types of HLP. The reduction in autoimmune shifts in CIHD patients during the treatment with an antiatherosclerotic diet including soya isolate, the simultaneous reduction in HLP and the drop in the activity of the kallikrein-kinin system and capillary permeability demonstrate the advisability of expanding the use of soya protein in the dietary treatment of CIHD patients. Increasing the overall amount of protein in an antiatherosclerotic diet — even with soya protein — partially levels off the therapeutic effect exerted by the antiatherosclerotic diet on hemacoagulation activity, however. This emphasizes the necessity to differentiate (in terms of qualitative and quantitative composition) approaches to the application of protein in diet treatments for CIHD patients, taking into account the type of HLP as well as the presence or absence of blood hypercoagulation.

Modifying the portion of protein in diets is playing an ever-increasing role in the dietary treatment and nutritional prevention of IHD, in particular with regard to correcting primarily type II hyperlipoproteinemia (HLP).[2,25]

Convincing data have been obtained from numrous experimental observations about a

TABLE 5
Chemical Composition of Experimental Diets Including Isolated Soy Protein

Parameter	Diets			
	Basal	2/3 Soy protein	1/3 soy protein	2/3 Soy protein
Total protein (g)	70.0	71.8	73.7	68.0
Soy protein (g)	0	52.0	26.6	34.0
Total fat (g)	100.4	75.0	74.5	73.5
Vegetable fat (g)	36.3	26.3	24.9	27.1
PUFA (% of calories)	6.7	5.2	5.1	5.5
Total carbohydrate (g)	433	436	422	425
Refined carbohydrate (g)	71	81	87	85
Cholesterol (mg)	330	280	300	370
Calories (kcal)	2910	2710	2650	2640

hypocholesterolemic effect and the deceleration in the development of atherosclerosis in animals when replacing all or part of the casein in atherogenic diets with soya protein.[22,55] Similar studies with human data also show a marked drop in the blood cholesterol level in both healthy individuals as well as in patients with HLP when soy protein is substituted with animal protein.[26] According to data obtained from various studies, the strength of the hypocholesterolemic effect exerted by the soya proteins fluctuates within wide limits, ranging from 7 to 30% of the initial level.[14,57] This apparently has to do with the heterogeneity of the groups examined, the different duration of each diet treatment, and the varying quantities of soya protein employed in the diets, which ranged from 13 to 92% of the overall caloric value of the diet.[17,58] Another factor at work is the difference in the content and quantity of other nutrients in the rations selected, such as cholesterol, animal fats, PUFA, food fibers, as well as in the calorific value. All of these factors may either moderate or potentiate the hypocholesterolemic effect of soya protein.[13]

The objective of these examinations was to determine the extent of the hypocholesterolemic effect exerted by soya protein given various amounts of the latter in three separate isocaloric rations identical in terms of the content and qualitative composition of other nutrients. Used as the source of soya protein were the following combined products containing 500E soya isolate (manufactured by Protein Technologies International, U.S.): soya beverage, oatmeal porridge, and noodle soup, all containing soya protein. Four successive diets were employed, each identical in chemical composition and caloric value and differing only in the qualitative composition of the protein. The background ration contained traditional sources of vegetable and animal protein in the Soviet diet, while two thirds of the overall quantity of protein (by weight) was replaced with a soy protein isolate in diet I, and one third and one half were substituted in diets II and III, respectively. The chemical composition of the diets is shown in Table 5.

The observations were conducted on 16 male IHD patients with type IIa HLP, ranging in age from 43 to 56 who had suffered a documented myocardial infarction 9 ± 5 years previously. Excess body weight (EBW) was observed in 15 of the 16 patients, with the average for the group totaling 76.7 ± 4.3 kg at a height of 169 ± 1.6 cm. All patients were given the background, I, II, and III diets in succession over four periods lasting 14 d each. In addition to the diets, the patients were also given 1 "Undevit" tablet, 0.5 calcium gluconate, and, as indicated, fast-acting or time-release nitrites. The observations were conducted under hospital conditions, with a controlled regime of physical exercise. Examinations were carried out on the day the patients signed into the hospital and on the last day of each period of observation. Blood samples were taken following a 14-h fast. The content of total cholesterol,[42] high-density lipoprotein (HDL) cholesterol,[58] and triglyceride[43] in the blood serum was determined, as were the activity of kallikrein and prekallikrein,[48]

TABLE 6

Change in the Indices for Lipid Metabolism in Blood Serum at Different Levels of Isolated Soy Protein in the Diet

Parameter	A Pretest	B Basal	C 2/3 Soy	D 1/3 Soy	E 1/2 Soy	Change (B — E)
Total cholesterol	7.56 ± 0.26	7.90 ± 0.36	6.73 ± 0.19	6.80 ± 0.1	5.98 ± 0.11	
Percent change[a] initial	—	± 4	-11^b	$+2$	-11^b	-24^b
HDL cholesterol	1.32 ± 0.03	1.42 ± 0.10	1.17 ± 0.03	1.15 ± 0.09	1.06 ± 0.02	
Percent change vs. initial[a]	—	$+8$	-18^b	-2	-8^b	-25^b
LDL cholesterol	6.01 ± 0.21	5.84 ± 0.18	5.11 ± 0.18	5.02 ± 0.07	4.54 ± 0.19	
Percent change vs. initial[a]	—	-3	-12^b	-2	-10^b	-22^b
Triglycerides	1.34 ± 0.05	1.51 ± 0.07	1.04 ± 0.06	1.44 ± 0.01	1.17 ± 0.08	
Percent change vs. initial[a]	—	$+13$	-31^b	$+38^b$	-19^b	-22^b

[a] Minus $(-)$ equals reduction; plus $(+)$ equals increase.

[b] Equal $p < 0.05$.

alpha$_2$-macroglobulin,[49] enzyme activity in the pancreas,[50] and the activity of cellular and humoral immunity.[19,20] The results obtained were statistically processed using the Student t test.

All of the patients (excluding one who refused to continue the diet therapy) came through the diet treatment in good shape; an unpleasant "chemical" taste in the mouth and gas were observed by only a few of the patients in the first 3 d on ration I, and both disappeared on their own. All patients quickly became tired of the diet, noting the monotony and bland taste, in particular when receiving diet I, even though this did not become a reason for discontinuing the diet therapy.

The adequate energy value in all rations notwithstanding, the patients with EBW lost weight uniformly throughout the entire period of observation (not statistically significant). Other authors have observed a reduction in weight given rations with a sufficient energy value in which high levels of animal protein were replaced with soya protein.[14,57] The changes observed in the indices for lipid metabolism in the blood serum during the diet treatment of IHD patients with type IIa HLP (Table 6) may not be explained by changes in body weight, however. There is a clear-cut connection between a reduction in the overall level of cholesterol in the blood serum in the patients examined and the quantity of soya protein, regardless of how much cholesterol and PUFA were in the rations employed. Replacing one third of the animal protein with isolated soya protein had virtually no hypocholesterolemic effect, which is apparent only at the higher one half and two thirds soya protein substitution levels. Overall blood cholesterol dropped to 24% of the initial level over the entire period of observation, and was more pronounced relative to the data which we obtained using diets rich in protein containing soya protein.[59]

In addition, a reduction in the atherogenic lipid fractions, low-density lipoprotein cholesterol, was found in the blood serum of the patients fed isolated soy protein diets. This drop was most pronounced during diet III (two thirds protein), and the atherogenic coefficient remained unchanged throughout remaining periods. Published literature contains contradictory data with regard to the influence exerted by soya protein on the content of HDL cholesterol. Noting the hypocholesterolemic effect of soya protein, several authors failed to observe changes in the level of HDL cholesterol,[17] while others write about a parallel reduction in cholesterol in all lipoprotein fractions.[33,60] This apparently has to do with the varying quantity of soya protein, the heterogeneity of the chemical composition of the rations, and the different level of their hypocholesterolemic effect. Literature data concerning the influence of soya protein on the level of triglycerides in the blood are ambiguous, as well.[28,33] In our examinations, the content of triglycerides in the blood serum exhibited an overall

TABLE 7

Change in the Activity of the Kallikrein-Kinin System of Pancreatic Enzymes in the Blood Serum Before (I) and After (II) Diet Treatment with Rations Containing Soya Protein Isolate

Indices examined	I Before	II After	Change in absolute units	Change in percent of initial value	Reliability of change
Kallikrein IE/ml	80.3 ± 12.8	130.8 ± 15.3	+50.5	+63%	$p < 0.05$
Prekallikrein IE/ml	190.2 ± 22.1	219.0 ± 25.3	+28.8	+15%	$p < 0.05$
L^2Macroglobulin IE/ml	1.14 ± 0.06	0.92 ± 0.1	−0.22	−19%	$p < 0.05$
L^1Trypsin inhibitor IE/ml	33.9 ± 6.6	34.4 ± 3.7	−0.5	+1.5%	$p < 0.05$
Ethyl benzoyl arginine UE/ml/min	0.232 ± 0.02	0.212 ± 0.015	−0.02	−9%	$p < 0.05$
Trypsin, med	3.27 ± 0.54	2.57 ± 0.28	−0.70	−21%	$p < 0.05$
Trypsin inhibitor, med	478 ± 19	524 + 10	+46	+10%	$p < 0.05$

tendency to decrease over the entire period of observation, undergoing significant changes as a function of the quantity of soya protein and independently of other dietary parameters and changes in body weight. The level of triglycerides in the blood was observed to increase in the case of the background diet and the diet containing the soya protein replacing one third of the protein. Other authors have noted the same tendency on the part of the triglycerides to increase given a reduction in the overall quantity of protein in the diet, particuarly in vegetarians.[2,33,61]

The mechanism underlying the hypocholesterolemic effect of soya protein is unclear. There are reports about an accelerated cholesterol metabolism in the liver,[19] elevated excretion of acidic and neutral sterols with the feces,[26] and reduced secretion of triglycerides and cholesterol by the liver in diets containing soya protein. Several researchers link the hypocholesterolemic effect of diets containing soya protein to the influence exerted by saponins[17] or other dietary components, i.e., phytosterol,[25] and to nonprotein factors contained in the soya protein isolate, which are present in amounts exceeding 100 g/kg.[60] One concept being advanced by other authors which would seem to rest on more solid ground attributes the antiatherogenic effect of some soya protein to the characteristic features of its amino acid composition, in particular to the elevated content of lysine and threonine and the low level of methionine. This hypothesis is borne out by the data which we obtained during an experiment concerning the damaging impact which methionine has on the endothelium and the involvement of threonine in the catabolism of cholesterol.[62] A change in the blood cholesterol level in the patients examined might be connected with the altered ratio between lysine and arginine in the rations employed; the drop in hypocholesterolemia was found to be more pronounced in rations I and II, in which the lysine to arginine ratio was 1:2, while measuring 0.8 and 0.9 in the background and II rations, respectively.[62] Soya protein is not deficient in any one amino acid, but by the imbalance of the amino acid composition of the diet, may in turn affect the digestion and metabolism of amino acids.[63] The reduced secretion by the kidney and lowered concentration of A1 apoproteins may have something to do with amino acid imbalances, as was noted in experiments on rats which were fed soya protein.[59] The most noticeable reduction in the content of HDL cholesterol in the blood serum in the patients which we examined in ration I (two thirds soya protein) might also be connected to a reduction in A1 apoproteins. The overall trypsin inhibitor alpha$_1$-antitrypsin (α-AT) in the blood serum showed a tendency to increase during the course of the diet treatment with diets including isolated soya protein, with a simultaneous tendency on the part of tryptic activity displayed by the blood and alpha$_2$-macroglobulin to decrease. This could have something to do with the presence of protein inhibitors in the isolated soya protein (Table 7), which help disrupt the secretion and suppression of the activity exhibited by trypsinase-

TABLE 8
Change in Indices for T Cell Immunity in IHD Patients During Diet Therapy

Parameter	Initial value	Period of evaluation[a]			Control
		I	II	III	
Lymphocyte blastotransformation reaction (%) under the influence of the following:					
PHA	53 ± 1.24	55.0 ± 1.62	58.0 ± 1.09	60.0 ± 1.02[b]	60 ± 6.0
KonA	55 ± 1.32	52.0 ± 1.29	59.0 ± 1.39	61.0 ± 1.02	67 ± 6.0
HTE	7 ± 0.8	5.4 ± 0.65	4.6 ± 0.36	4.0 ± 0.37	6 ± 0.2
AAE	7 ± 0.73	4.6 ± 0.54	4.7 ± 0.58	4.0 ± 0.51	5.5 ± 0.4
Index of inhibitions of leukocyte migration in the presence of the following:					
PHA	0.79 ± 0.01	0.80 ± 0.02	0.84 ± 0.01	0.86 ± 0.01[b]	0.85 ± 0.03
HTE	0.72 ± 0.01	0.72 ± 0.02	0.80 ± 0.01	0.84 ± 0.01[b]	0.85 ± 0.03
AAE	0.71 ± 0.01	0.75 ± 0.02	0.80 ± 0.01	0.85 ± 0.01[b]	0.85 ± 0.03
Index of cytotoxicity of lymphocytes in the presence of the following:					
PHA	0.41 ± 0.01	0.39 ± 0.01	0.35 ± 0.01	0.33 ± 0.01[b]	0.35 ± 0.01
HTE	0.23 ± 0.01	0.21 ± 0.01	0.20 ± 0.01	0.17 ± 0.01[b]	0.15 ± 0.01
AAE	0.24 ± 0.01	0.21 ± 0.01	0.20 ± 0.01	0.17 ± 0.01[b]	0.15 ± 0.01

Note: HTE, water-salt heart tissue extract; AAE, water-salt extract from atherosclerotically altered aorta; PHA, phytohemagglutinin; KonA, concanavalin A.

[a] I = two thirds of protein from isolated soy protein; II = one third protein from isolated soy protein; III = one half of protein from isolated soy protein.

[b] $p < 0.05$.

type serine proteinases.[33] At the same time, the activity of kallikrein increased significantly, while prekallikrein revealed a trend toward elevated activity. As pointed out in our preceding publications,[60] replacing a portion of the animal protein with an isolated soya protein (15%) in a diet containing a high level of protein (120 g) actually helped to decrease the activity of the kallikrein-kinin system in the blood.

The complicated set of symptoms surrounding IHD is determined not only by the various combinations of metabolic disruptions closely linked to the kind of diet, but also by the involvement of cell-mediated immune reactions in the pathogenesis of the disease.

The immunopathological shifts discovered during the phase in which the symptoms of the disease have already taken shape later have a significant impact on the course taken by the illness, determining to a large extent whether or not it will be chronic and progressive in nature.[62,63]

We have shown earlier the positive effect of diet treatment on immunological changes which accompany CIHD. Positive changes occurred in the indices for cell immunity in most of the patients during the diet treatment with diets which included soya proteins (Table 8). Of particular interest was the discovery of an increase in the sensitivity of the lymphocytes to polyclonal mitogens (phytohemagglutinin and concanavalin A) accompanied by a simultaneous drop in the activity exhibited by mononuclears in response to specific antigens (water-salt extract from tissue of the heart and aorta damaged by the atherosclerotic process). A significant increase in the initially lowered levels of a third component of the complement was observed as well as a statistically significant reduction in the orosomucoid titers (Table 9). The level of immunoglubulins in the blood serum remained unchanged following diet therapy.

In summarizing the above experiment, one may state that future observations will in all likelihood make it possible to more precisely define requirements relating to the level of soya protein that should be included in food rations for IHD patients. This would ensure the biological and nutritional value of the rations to the greatest extent possible, and have a corrective influence on the altered metabolism of cholesterol.

TABLE 9
Changes in the Protein Components in Blood Serum of IHD Patients During Dietary Treatment

| Parameter | Initial value | Period of evaluation | | | Control |
		I	II	III	
Immunoglobulins (mg%)					
C	1343.0 ± 22.6	1340.0 ± 47.0	1389.0 ± 31.0	1379.0 ± 31.0	1268.0 ± 30.0
M	145.0 ± 3.0	139.0 ± 6.0	141.0 ± 3.0	141.0 ± 4.0	143.0 ± 10.0
A	217.0 ± 4.0	188.0 ± 4.0	189.0 ± 4.0	185.0 ± 3.0	225.0 ± 14.0
D	4.8 ± 0.15	4.4 ± 0.25	4.7 ± 0.17	4.4 ± 0.12	3.3 ± 0.2
Complementary fractions (mg%)					
C 1 Inhibitor	37.0 ± 0.95	29.0 ± 0.86	44 ± 0.66	37.0 ± 0.66	35 ± 2.3
C 3	116.0 ± 2.5	117.0 ± 2.8	125 ± 1.0	128.0 ± 1.0^a	132 ± 8.5
C 4	28.0 ± 0.95	27.0 ± 0.54	29 ± 0.73	29.0 ± 0.37	30 ± 2.0
C 5	9.0 ± 0.73	9.0 ± 0.54	8 ± 0.30	10.0 ± 0.44	10 ± 0.3
C 9	11.5 ± 0.51	9.4 ± 0.76	10 ± 0.37	11.6 ± 0.22	12 ± 0.2

Note: I = two thirds of protein from isolated soy protein; II = one third protein from isolated soy protein; III = one half of protein from isolated soy protein.

[a] $p < 0.05$, cell groups compared with initial values.

REFERENCES

1. **Pokrovskii, A. A.,** The role of nutrition in the prevention of disease associated with disruptions im metabolism, *Ter. Arkh.,* 1, 23, 1974.
2. **Samsonov, M. A. et al.,** The importance of protein quotas in the diet of persons suffering from ischemic heart disease, *Vestn. Akad. Med. Nauk SSSR,* 10, 68, 1978.
3. **Masek, Y.,** Nutrition and obesity, *Vnitr. Lek.,* 7, 736, 1961.
4. **McCracen, B. H.,** Etiological aspects of obesity, *Am. J. Med. Sci.,* 243, 99, 1962.
5. **Anderson, Y.,** Obesity, *Br. Med. J.,* 1 N5799, 560, 1972.
6. **Kajaba, I. and Grunt, Y.,** Kotazke kriterii pre posudenic vyskytu obezity u populaenych celkov, *Cesk. Gastroenterol. Vyz.,* 28, 352, 1972.
7. **Marks, H. H.,** Influence of obesity on morbidity and mortality, *Bull. N.Y. Acad. Med.,* 36, 296, 1960.
8. **Beyul, E. A. and Popova, Yu. P.,** Obesity as a social problem of modern times, *Ter. Arkh.,* 1, 71, 1984.
9. **Beyul, E. A., Oleneva, V. A. et al.,** Ways of determining and preventing obesity, *Vopr. Pitan.,* 4, 22, 1980.
10. **Frolova, I. A. and Oleneva, V. A.,** Several characteristics of lipid and catecholamine metabolism during the complication of alimentary obesity by sugar diabetes in its latent form, *Vopr. Pitan.,* 4, 7, 1981.
11. **Samsonov, M. A., Meshcheryakova, V. A. et al.,** The influence exerted by fructose in anti-atherosclerotic diets on the course of ischemic heart disease, *Vopr. Pitan.,* 6, 17, 1984.
12. **Kaznacheev, V. P.,** *Primary Enzymatic Processes in the Pathology and Clinical Picture of Rheumatism,* Novosibirsh, 1969, 110.
13. **Carroll, K. K. and Hamilton, R. M.,** Effects of dietary protein and carbohydrate on plasma cholesterol levels in relation to atherosclerosis, *J. Food Sci.,* 40, 18, 1975.
14. **Carroll, K. K., Giovanetti, P. M., Huff, M. W. et al.,** Hypocholesterolemic effect of substituting soybean protein for animal protein in the diet of healthy young women, *Am. J. Clin. Nutr.,* 31, 1312, 1978.
15. **Carroll, K. K.,** Dietary protein in relation to plasma cholesterol levels and atherosclerosis, *Nutr. Rev.,* 1 N36, 1, 1978.
16. **Connor, W. W. and Connor, S. L.,** The key role of nutritional factors in the prevention of coronary heart disease, *Prev. Med.,* 1 N 1-2, 49, 1972.
17. **Huff, M. W., Hamilton, R. M. et al.,** Plasma cholesterol levels in rabbits fed lowfat cholesterol free, semipurified diets: effects of dietary proteins, protein hydrolysates and amino acid mixtures, *Atherosclerosis,* 28, 187, 1977.

18. **Jerushalmy, J. and Hilleboe, H. E.,** Fat in the diet and mortality from heart disease, a methodologic note, *N.Y. State J. Med.,* 57, 2343, 1957.

19. **Mancini, G., Carbonara, A. O., and Heremans, J. F.,** Immunochemical quantitation of antigens by single radial immunodiffusion, *Immunochemistry,* 2, 235, 1965.

20. **Roseanu, W. and Moon, H. D.,** Lysis of homologaus cells by sensitized lymphocytes in tissue culture, *J. Natl. Cancer Inst.,* 27, 471, 1961.

21. **Hamilton, R. M. G. and Carroll, K. K.,** Plasma cholesterol levels in rabbits fed low fat, low cholesterol diets; effects of dietary proteins, carbohydrates and fiber from different sources, *Atherosclerosis,* 24, 47, 1976.

22. **Huff, M. W. and Carroll, K. K.,** Effects on dietary and amino acid mixtures on plasma cholesterol levels in rabbits, *J. Nutr.,* 110, 1676, 1980.

23. **Anichkov, N. N. and Tsinzerling, V. D.,** The problem of atherosclerosis as it stands today, in *Atherosclerosis,* Medgiz, Moscow, 1953, 7.

24. **Kritchevsky, D., Tepper, S. A., Czarnecki, S. K. et al.,** Experimental atherosclerosis in rabbits fed cholesterol-free diets, *Atherosclerosis,* 39, 169, 1981.

25. **Sirtori, C. R. et al.,** Soybean protein diet in the treatment of type II hyperlipoproteinemia, *Lancet,* 1, 275, 1977.

26. **Sirtori, C. R. et al.,** Clinical experience with the soybean protein diet in the treatment of hypercholesterolemia, *Am. J. Clin. Nutr.,* 39, 1645, 1979.

27. **Descovich, G. C. et al.,** Multicentre study of soybean protein diet for outpatient hypercholesterolaemic patients, *Lancet,* II, 709, 1980.

28. **Holmes, W. L. et al.,** Comparison of the dietary meat versus dietary soybean protein in plasma lipids of hyperlipidemic individuals, *Atherosclerosis,* 36, 379, 1980.

29. **Howard, A. N., Gresham, G. A., Jones, D. et al.,** The prevention of rabbit atherosclerosis by soya bean meal, *J. Atheroscler. Res.,* 5, 330, 1965.

30. **Kritchevsky, D.,** Diet and cholesteremia, *Lipids,* 12, 49, 1976.

31. **Kritchevsky, D., Tepper, S. A., Williams, D. E., and Story, J. A.,** Experimental atherosclerosis in rabbits fed cholesterol free diets, *Atherosclerosis,* 26, 403, 1977.

32. **Noseda, G., Frangiacomo, C., Bosia, C. et al.,** Behandlung der Hyperlipidamic typ II mit Sojabohnen, *Schweiz. Med. Wochenschr.,* Bd 109, N47, 1852, 1979.

33. **Shorey, R. A. L. et al.,** Determinants of hypocholesterolemic response to soy and animal protein-based diets, *Am. J Clin. Nutr.,* 34, 1769, 1981.

34. **Van Raaij, I. M. A. et al.,** Effects of casein versus soy protein diets on serum cholesterol and lipoproteins in young healthy volunteers, *Am. J. Clin. Nutr.,* 34, 1261, 1981.

35. **Kritchevsky, D., Tepper, S. A., and Story, J. A.,** Influence of soy protein and casein on atherosclerosis in rabbits, *Fed. Proc., Fed. Am. Soc. Exp. Biol.,* 37, 2801, 1978.

36. **Kritchevsky, D.,** Vegetable protein and atherosclerosis, *J. Am. Oil Chem. Soc.,* 56, 135, 1979.

37. **Olson, R. E., Nichaman, M. Z., Nittka, J., and Eagles, J. A.,** Effect of amino acid diet upon serum lipids in man, *Am. J. Clin. Nutr.,* 23, 1614, 1970.

38. **Torre, Y. M., Lynch, V., and Yarowski, C. I.,** Lowering of serum cholesterol and triglyceride levels by balancing amino acid intake in the white rat, *J. Nutr.,* 110, 1194, 1980.

39. **Malinow, M. R., McLaughlin, P., Kohler, Y. O., and Livingston, A. L.,** Prevention of elevated cholesterolemia in monkeys by alfalfa saponins, *Steroids,* 29, 105, 1977.

40. **Oakenfull, D. G. and Fenwick, D. E.,** Absorption of bile salts from aqueous solution by plant fiber and cholestyramine, *Br. J. Nutr.,* 40, 299, 1978.

41. **Potter, J. D., Topping, D., and Oakenfull, D.,** Soya, saponins and plasma-cholesterol, *Lancet,* 1, N 819, 223, 1979.

42. **Ilk, Z.,** Fast method for determining the content of overall cholesterol in the blood serum (according to Ilk), in *Biochemical Methods of Clinical Examination,* Pokrovsky, A. A., Ed., Moscow, 302.

43. **Neri, B. and Frings, C. S.,** Improved method for determination of triglycerides in serum, *Clin. Chem.,* 19, 1201, 1973.

44. **Laszar, Y.,** Determination of fibrinogen (and fibrinolysis) in small quantities of plasma, *Thromb. Diath. Haemorrh.,* 17, 401, 1967.

45. **Kowalski, E., Kopec, M., and Niewiarowski, S.,** An evaluation of the euglobulin method for the determination of fibrinolysis, *J. Clin. Pathol.,* 12, 215, 1959.

46. **Astrup, T. and Mullertz, S.,** Fibrin plate method for estimating fibrinolytic activity, *Arch. Biochem.,* 40, 346, 1952.

47. **Andreenko, G. V.,** Micromethods for determining the blood concentration of fibrinogen, fibrinolytic activity, heparin and antifibrinolysin, *Lab. Delo,* 8, 477, 1969.

48. **Paskhina, T. S. et al.,** Simplified method for determining kallikreinogen and kallikrein in human blood serum under normal conditions and given several pathological conditions, *Vopr. Med. Khim.,* 6, 660, 1974.

49. **Nartikova, V. F. et al.,** Unified method for determining the activity of antitrypsin and macroglobulin in human blood serum, *Vopr. Med. Khim.,* 4, 494, 1979.

50. **Khechinashvili, G. G. et al.,** Lipoprotein lipase, hepatic triglyceride lipase and several other lipolytic enzymes of the animal organism, *Usp. Biol. Khim.,* 21, 163, 1980.

51. **Erlanger, B. F., Kokowsky, N., and Cohen, W.,** The preparation and properties of two new chromogenic substracts of trypsin, *Arch. Biochem.,* 95, 271, 1961.

52. **Azimov, F. I. et al.,** The condition of the kallikrein-kinin blood system in patients who suffered myocardial infarctions and its reaction to physical exertion, *Kardiologiya,* 9, 54, 1983.

53. **Gomazkov, O. A.,** The kallikrein-kinin blood system and the regulation of hemodynamics, *Kardiologiya,* 7, 130, 1973.

54. **Murashko, V. V. et al.,** The activity of the kallikrein-kinin blood system in victims of coronary atherosclerosis, *Ter. Arkh.,* 4, 11, 1981.

55. **Huff, M. W. and Carroll, K. K.,** Effects of dietary protein on plasma cholesterol levels and cholesterol oxidation in rabbits, *Fed. Proc., Fed. Am. Soc. Exp. Biol.,* 36, 1104, 1977.

56. **Reiser, R., Henderson, G. R., O'Brien, B. C., and Thomas, Y.,** Hepatic 3-hydroxy-3-methyl-glutary coenzyme-A reductase of rats fed semipurified and stock diets, *J. Nutr.,* 107, 453, 1977.

57. **Terpstra, G. et al.,** The hypocholesterolemic effect of dietary soy protein in rats, *J. Nutr.,* 112, 810, 1982.

58. **Klimov, A. N. and Ganelina, I. E.,** *Phenotyping Hyperlipoprotienemia, Methodological Recommendations,* Moscow, 1975.

59. **Meshcheryakova, V. A., Samsonov, M. A. et al.,** The importance of the qualitative composition of proteins in the diet treatment of patients with ischemic heart disease, *Vopr. Pitan.,* 6, 3, 1985.

60. **Nagat, I. et al.,** Effects of soya-bean protein and casein on serum cholesterol levels in rats, *Br. J. Nutr.,* 44, 113, 1980.

61. **Burslem, J., Schonfeld, Y., Howald, M. A. et al.,** Plasma apoprotein and lipoprotein lipid levels in vegetarian, *Metab. Clin. Exp.,* 27, 711, 1978.

62. **Selivanov, I. I. and Khardaev, E. K.,** Blastotransformation of lymphocytes and the inhibition of leukocyte migration as sensitization indices during atherosclerosis of the coronary arteries of the heart, *Sov. Med.,* 1, 35, 1979.

63. **Osipov, S. G.,** Circulatory immune complexes and immune reactivity in individuals with ischemic heart disease, *Immunologiya,* 70, 1982.

64. **Fredrickson, D. S. and Levy, R. I.,** Familial hyperlipoproteinemia, in *The Metabolic Basis of Inherited Disease,* Stanbury, J. B., Wyngaarden, J. B., and Fredrickson, D. S., Eds., New York, 1972, 493.

Chapter 15

CURRENT DIETARY COMPOSITION IN DEVELOPED COUNTRIES AND NEED FOR MODIFICATION

Fred H. Steinke and Michael N. Volgarev

The dietary factors which relate to the development of heart disese, obesity, cancer, and diabetes have been reviewed and evaluated by numerous scientific committees over the last 30 years. The dietary components[1] which are consumed in excess and contribute to the development of these diseases or whose reduction will moderate their occurrence are

- excess calorie intake
- excess fat intake
- excess saturated fat intake
- high salt intake
- high cholesterol intake

The primary sources of fat and saturated fats are animal products. These include meat, meat products, dairy products, and eggs. The fats in animal products contain a large amount of saturated fat as well as large amounts of total fat and a high percentage of calories provided by fat. Animal products also are the sole source of cholesterol in the diet with eggs being the largest concentrated source in the diet.

A reduction of 1% of calories from saturated fat has been estimated by Keys and Parlin[2] and Hegsted et al.[3] to reduce the serum cholesterol by 2.7 mg/dl. Therefore, reductions in animal fat which have a high saturated fat content will have a marked reduction in serum cholesterol and risk of heart disease. A 10% reduction in calories from saturated fat would result in a 27-mg reduction in serum cholesterol. For an individual with a cholesterol level of 200 mg/dl, this would reduce the risk of heart attack by 27% based on the 2:1 ratio of heart attack on serum cholesterol or 18% in an individual with a cholesterol level of 300 mg/dl.

The reduction in animal fat sources is important in reducing not only the saturated fat, but specifically the C:14 (myristic) and C:16 (palmitic) fatty acids, which have been identified as being hypercholesterolemic. Butter fat and products containing butter fat, such as milk and cheese, are particularly high in both total saturated fats and the C:14/C:16 fatty acids. Most cheeses contain in excess of 70% of their calories from fat and more than 40% of calories from saturated fat.

The use of these foods in the diet has been traditional and developed into the eating customs of the populations over centuries. The taste and acceptability of these products are high and therefore preferred by the consumer. Therefore, in order to change the consumption of the foods, it will be necessary to provide new foods which have equal or superior taste and flavor without the undesirable fat and cholesterol content. At the same time, these traditional foods have been sources of important nutrients in the diet which need to be maintained in the new foods and eating patterns.

In order to modify the diet or food products available to the consumer it is first necessary to know the magnitude of the dietary changes needed and the present status of the diet patterns. The general recommendation of numerous scientific committees[1,5,6] is to reduce the amount of fat consumed to less than 30% of calories, saturated fat to less than 10% of calories, and cholesterol to less than 300 mg/d.

The fat consumption has been estimated to be in excess of 40% of calories in most

TABLE 1
Calories, Fat, and Cholesterol
Intakes in Sample Populations

	U.S.	U.S.S.R.
Calories		
kcal/d	2554	2568
kcal/kg body weights	32	35
Fat		
g/d	116	110
Percent of calories	40	38
Saturated fat		
g/d	43	47
Percent of calories	15	16
Cholesterol		
mg/d	468	508

Summarized from *Am. J. Clin. Nutr.*, 39, 942, 1984.

TABLE 2
Amounts of Dietary Fat at 30 and 40% of Calories for Various Calorie Intakes

	2200 kcal	2500 kcal g fat	3000 kcal
Fat, total			
40% of calories	98	111	133
30% of calories	73	83	100
Difference in g of fat per day	25	28	33
Saturated fat			
17% of calories	41	47	57
10% of calories	24	28	33
Difference in grams of saturated fat per day	17	19	24

developed countries. This was confirmed by survey data reported in a joint U.S.-U.S.S.R. project.[4] The summary of the results (Table 1) showed similar daily calorie intake of 2554 and 2568 kcal for the U.S. and U.S.S.R., respectively, by adult males. Fat consumption was high in both countries at 40 and 38% of total calories, and saturated fat accounted for 15 and 16% of calories. The upper 10% of the population in these surveys were consuming more than 50% of their daily calories from fat, and saturated fat accounted for more than 20% of calories.

The use of percent of calories as a way of evaluating food composition and food consumption is difficult for the consumer and food producer to comprehend and use in their normal life-style. The percentage of calories at 30 and 40% of dietary calories has been converted to specific amounts of fat in Table 2 for calorie intakes of 2200, 2500, and 3000 kcal/d. The 2200 kcal/d would be applicable to women, while 2500 and 3000 kcal are applicable to men, depending on their size and activity level. The amounts of fat which need to be removed from the diet in order to move the average fat intake from 40 to 30% of calories ranges from 25 to 33 g of fat. Saturated fat needs to be reduced by 17 to 24 g to reduce the saturated fat to 10% of calories at calorie intakes of 2200 to 3000 kcal/d. The cholesterol intakes typically average 500 mg/d, where the recommended intakes are less than 300 mg/d.

The sources of this excess fat intake are high-fat meats, dairy products, and cooking

TABLE 3
Distribution of Sausages and Luncheon Meats by Fat Content

	U.S.S.R. sausage		U.S.A. sausage and luncheon meats	
Percent fat in product	Number	Percent of total	Number	Percent of total
0 to 15	3	5	21	27
15 to 25	25	39	24	30
25 to 35	13	20	29	37
35 to 45	16	25	5	6
45 +	7	11	0	0

fats. Normal meat cuts allow the removal of excess fat through trimming during food preparation and eating. However, much of the meat consumed does not allow for separation of fat during preparation or cooking. Most processed meats fall into this category, and can amount to large portions of the meat and fat intake. In the U.S., large quantities of ground beef are consumed, which contains 20 to 30% fat and has 70 to 80% of calories from fat. In the U.S.S.R., large percentages of meat are consumed as sausages of a wide variety. Similar types of sausages and processed meats are also consumed in the U.S. for breakfast, lunch, and dinner.

The distribution of these sausages and luncheon meats based on nutritional composition tables are summarized in Table 3. The sausages and luncheon meats available in the U.S.S.R. tend to be high in fat, and 56% of the types of products available have fat contents of 25% or greater. Only three products (or 5%) had fat content less than 15%. The products containing 25% or more fat in the U.S. (U.S. Department of Agriculture Agricultural Handbook 8-8) accounts for 43% of the total products. The Soviet products do have much larger numbers of products in the highest fat content categories. The high percentage of lowest fat category in the U.S. products is in part due to the larger use of chicken and turkey products which are inherently lower in fat content. These distributions, of course, do not necessarily correlate to consumption of fat, since data are not available on production or purchasing of these food types by consumers.

However, reduction in fat content of sausages and luncheon meats by 50% will provide a significant improvement in the diet. A reduction in the fat content from 40 to 20 g/100-g serving will result in 20 g less fat. If these types of meats are consumed twice daily, this would be equal to 40 g less fat consumed per day. This quantity of reduction would exceed that necessary to reduce the dietary calories from fat by 10% of calories from 40 to 30% of calories. Even greater reduction can be obtained by replacing these high-fat meats totally with lower-fat, nonmeat foods such as breakfast cereals and porridges.

The advantage of modifying existing food products ensures rapid consumer acceptance and reduces the cost and time required to conduct nutrition education programs that retrain the general population in eating patterns.

Dairy products also constitute a large potential source of dietary fat and saturated fat. In addition, milk protein has reported hypercholesterolemic effects in some people, and its reduction may provide additional benefits to these people. The simplest method of reducing dairy fat is to reduce the amount of butter consumed. The amount of butter which cannot be reduced in the diet would be replaced by margarine and vegetable cooking oils, which contains less saturated fatty acids. The reduction in butter consumption has in fact taken place on a large scale in the U.S. during the last 40 years, with the much wider use of margarine and vegetable oils.

Full-fat milk also provides significant amounts of fat to the diet. A single 250-ml serving (a glass) provides 8 g of butter fat. Two glasses per day will provide 16 g. Thus, modifying or reducing the amount of milk consumed will have a significant effect on both total and saturated fat intake. This has in fact occurred in the U.S. The consumer education programs have resulted in a marked decrease in full-fat milk and greater consumption of 2% fat milk and skim milk.

Many cheeses are also high in fat and percent of calories from fat. These include cheddar, cream cheese, Gruyere, and sour cream types. These contain 70 to 90% of calories from fat. Regular ice cream is also high in fat (61% of calories). Changing to lower-fat cheeses and desserts should be a goal of food processing.

The food production over the centuries has developed a wide variety of highly tasty food products that have high consumer appeal. Unfortunately, many of these products are also very high in fat and calories. To improve the health of the population, these foods need to be modified to reduce their undesirable high-fat content while maintaining their high acceptability. In coordination with this change in direction, a strong education program is needed to make the consumer aware of which products are healthy.

REFERENCES

1. Prevention of Coronary Heart Disease, Report of a WHO Expert Committee, Tech. Rep. Ser. No. 678, World Health Organization, Geneva, 1982.
2. **Keys, A. and Parlin, R. W.,** Serum cholesterol response to changes in dietary lipids, *Am. J. Clin. Nutr.,* 19, 175, 1966.
3. **Hegsted, D. M., McGandy, R. B., Meyers, M. L., and Stare, F. J.,** Quantitative effects of dietary fat on serum cholesterol in man, *Am. J. Clin. Nutr.,* 17, 281, 1965.
4. Nutrient intake and its association with high density lipoprotein and low density lipoprotein cholesterol in selected U.S. and USSR sub-populations; the US-USSR Steering Committee for Problem Area 1: the pathogenesis of atherosclerosis, *Am. J. Clin. Nutr.,* 39, 942, 1984.
5. Lowering blood cholesterol to prevent heart disease, *J. Am. Med. Assoc.,* 253, 2080, 1985.
6. *Diet and Health, Implications for Reducing Chronic Disease Risk,* National Research Council, National Academy Press, Washington, D.C., 1989.

Chapter 16

PRODUCTION OF NEW, HEALTHIER FOODS USING ISOLATED SOY PROTEIN

Fred H. Steinke, Stanley H. Richert, and Charles W. Kolar

TABLE OF CONTENTS

I. FORMULATION OF LOW-FAT PROCESSED MEAT PRODUCTS

The consumption of large amounts of meat, egg, and milk products has been identified as contributing to the rise in the incidence of heart disease during the last 100 years.[1] These foods have increased the intake of saturated fat, total fat, and cholesterol of people in developed countries. In addition, the reduction in physical exercise by many people has resulted in a decline in the need for calories. These combined factors have resulted in a tendency to overconsume calories, thus depositing excess body fat. Overweight has been associated with increased occurrence of heart disease, diabetes, and some cancers.[2]

Obesity itself is associated with early mortality. The costs of these diseases to the nation in medical costs and lost productivity are enormous. The annual medical costs of heart disease are estimated to be $85 billion in the U.S. alone. The costs to individuals developing these chronic diseases and their families are of course even greater, both financially and emotionally.

The World Health Organization,[1] as well as many national[2-5] and regional health organizations, have identified and recommended changes in dietary patterns and physical activity to reduce the risks of developing heart disease, diabetes, cancer, and obesity. These recommended changes specifically call for a reduction in the consumption of saturated fat, total fat, and cholesterol by reducing the intake of fats, meats, dairy products, and eggs. Conversely, these oganizations recommend an increase in vegetable products which do not contain saturated fat and are lower in calorie content.

This change can be accomplished by massive education programs and changes in eating habits. However, many consumers will resist change, especially to less-palatable foods. The consumer has selected the present food consumption patterns because it was highly acceptable. Therefore, an alternative to massive dietary changes is to modify foods which the consumer already likes and consumes to a more desirable composition. Specifically, animal products high in fat and cholesterol can be modified to be lower in fat and cholesterol while retaining their high acceptability by the consumer.

The modifications in existing foods can in part be done by careful selection and trimming of meat fat, and separating and discarding butter fat. However, it is difficult to remove cholesterol from eggs or meat. These changes will result in a higher cost of foods, and in some societies a limitation on available food. An alternative is to use new protein foods to replace and dilute the animal products while maintaining a highly nutritious profile and economical product. Simple reduction in fat content often results in decreased palatability in the food caused by increased toughness and decreased juiciness. These effects can often be avoided and product acceptability maintained by using new protein sources and newer processing technology.

The reductions in dietary fat and cholesterol recommended by the World Health Organization,[1] National Institutes of Health,[2,3] and the American Heart Association[4] are summarized in Table 1. The changes are given in percentage of the calories consumed to standardize the recommendations across the variable energy requirements by age and sex. The target level for total fat intake for the general population is to consume less than 30% of calories from fat.

Since the typical consumption of fat in the diet is 40% or more of calories,[4] this means that the typical individual needs to reduce their fat intake by 10% of their daily caloric intake. The calorie requirement of the typical adult male is estimated to be 3000 kcal and the adult female 2200 kcal per day. This equates to a needed change of 300 and 220 kcal of fat per day or a reduction of 33 and 24 g of dietary fat per day for males and females, respectively. Since food products are purchased on the basis of weight rather than percent of calories, these latter numbers in grams of fat are more easily understood by the food

TABLE 1
Dietary Changes Recommended to Reduce Risk of Heart Disease

Food component	Typical diets	Recommended diet composition	Changes needed	Amount of reduction needed to meet goals[a]	
				3000 kcal[b]	2200 kcal[c]
Fat, total	40%+ of kcal	30% of kcal	−10% of kcal	33 g	24 g
Fat, saturated	17% of kcal	10% of kcal	−7% of kcal	23 g	17 g
Cholesterol	500 mg	300 mg	−200 mg	200 mg	200 mg

[a] Kilocalories converted to grams using the factor of 9 kcal/g of fat.
[b] Kilocalorie requirement for the adult male (Tech. Rep. Ser. No. 724, World Health Organization, 1985).
[c] Kilocalorie requirement for the adult female (Tech. Rep. Ser. No. 724, World Health Organization, 1985).

processor and consumer. The typical consumption of saturated fat is approximately 17% of calories, while the target goal is to reduce this to less than 10% of calories. This equates to 23 and 17 g of saturated fat for males and females, respectively.

In addition to fat reduction, dietary cholesterol reduction has also been recommended for both the general population and individuals with hypercholesterolemia. The cholesterol intake varies greatly, but is on average about 500 mg/d in developed countries. The target is to reduce this level below 300 mg/d for the general population and even lower for individuals with blood cholesterol levels above 240 mg. The primary sources of cholesterol in the diet are eggs, which contribute approximately 270 mg per egg to the diet. However, meat and dairy products do add some cholesterol to the diet, and changes in their consumption will add to the reduction in the daily intake of cholesterol. Eggs are often found in processed foods, including processed meats. The replacement of these eggs in processed foods can contribute significantly to the reduction in cholesterol intake.

The amount of fat and saturated fat in processed meats can be reduced by selecting leaner meats and reducing the amount of fat incorporated into the formulations. The fat content of a typical frankfurter is 30%. This could be reduced to 15 or 20% by carefully selecting the meats. However, this would result in a change in the texture and appearance, therefore reducing the acceptability of the product. Figure 1 shows the effects of fat levels on textural strength as measured by the Instron® equipment. The textural strength increased in a linear fashion as the fat content decreased. In addition, taste panel evaluation also confirmed a perceptual increase in hardness and chewiness when the fat content was reduced from 30 to 15% in all meat frankfurters. Frankfurters made with 15% fat with hydrated isolated soy protein replacing about 25% of the meat still maintained the organoleptic characteristics of the traditional frankfurter (see Table 2). Thus, the addition of a highly functional soy protein maintained the preferred textural characteristics, as well as provided a nutritionally adequate protein source for the replacement of the meat protein.

The composition of a typical frankfurter formulation and a modified formulation using an isolated soy protein (PP500E) are given in Table 3. The new formulation with isolated soy protein resulted in a product with a higher protein content, as well as having 50% less fat, 40% fewer calories, and 34% less cholesterol. This demonstrates the capability of modifying processed meats, which make a substantial contribution to the diet in many developed countries.

Wide varieties of processed meat products are available and consumed at all three meals. Table 4 gives a partial list of the processed meat products available in the U.S. with their fat, calorie, and cholesterol content. Each country has its own unique products and name identification. However, the data in the table illustrates the typical composition which can be expected in processed meats. The fat content of these products generally ranges from 15 to 40%. The caloric content varies with the fat content. The cholesterol content, however,

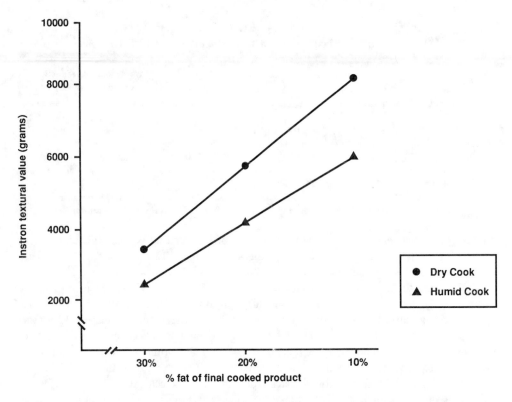

FIGURE 1. Effects of lowering the fat content upon the Instron® texture of a frankfurter cooked using a dry schedule and a humid schedule.

TABLE 2
Sensory Evaluation of Frankfurter Models

Sensory parameter	Traditional (30% fat)	All meat (15% fat)	Isolated soy protein/meat (15% fat)
Cured meat color	8.06 a	11.31 b	7.21 a
Texture			
Hardness	7.24 a	11.64 b	6.94 a
Juiciness	10.54 a	8.64 a	8.60 a
Chewiness	5.65 a	7.41 b	6.11 a
Flavor			
Meatiness	8.50 b	9.88 b	7.56 a
Spiciness	7.52 a	8.69 a	6.71 a
Aftertaste	4.50 a	5.06 a	5.00 a

Note: 0 equals pale, soft, dry, lack of, or weak, depending on the sensory parameter. 15 equals the maximum in intensity, hardness, juiciness, number of chews, or strength. Rows containing the same letter are not significantly different ($p < 0.05$).

does not necessarily depend upon the fat content, but is related to the type of meat products included in the formulation. Products with organ, skin, and nerve tissue tend to be higher in cholesterol content than those made from muscle meat. Although chicken products are normally believed to be lower in fat and cholesterol, chicken frankfurters contain 19.5 g of fat and 101 mg of cholesterol ner 100 g. While the fat content is lower than the standard

TABLE 3
Comparison of Frankfurter Formulas

Ingredient	Unit	Traditional product	Low-fat isolated soy protein/meat product
Meat blend (38% fat)	%	71.6	—
Meat blend (27% fat)	%	—	51.3
Nonfat dry milk	%	1.5	1.5
Whole egg	%	2.3	—
PP500E	%	—	3.8
Salt and spices	%	2.0	2.0
Water	%	22.6	41.4
Total	**%**	**100.0**	**100.0**

Composition	Unit	Amt per 100 g	Amt per 100 g	Percent change
Protein	g	12.1	14.1	+16
Fat, total	g	30.0	15.0	−50
Fat, saturated	g	11.3	5.8	−49
Calories	kcal	322.0	195.0	−39
Cholesterol	mg	65.0	43.0	−34

TABLE 4
Fat, Calorie, and Cholesterol Content of Typical Processed Meat Products

Processed meat	Amount per 100 g Fat (g)	Calories (kcal)	Cholesterol (mg)
Beerwurst, beef	29.4	324	56
Berliner, beef and pork	17.2	230	46
Blood sausage	34.5	378	120
Bologna, beef and pork	28.3	316	55
Bratwurst, pork	25.9	301	60
Braunschweiger	32.1	359	156
Brotwurst	14.3	323	63
Dutch brand loaf	17.8	240	47
Frankfurter, beef and pork	29.2	320	50
Frankfurter, chicken	19.5	257	101
Ham, chopped	18.8	239	49
Italian sausage, raw	31.3	346	76
Kielbasa	27.2	310	67
Knackwurst	27.8	308	58
Lebanon bologna	14.8	226	66
Luncheon meat, pork, beef	32.2	353	55
Luncheon sausage	20.9	260	64
Mortadella	25.4	311	56
Olive loaf	16.5	235	38
Pate, goose	43.8	462	150
Peppered loaf	17.3	148	46
Pork sausage, raw	40.3	417	68
Salami, dry, beef and pork	34.4	418	79
Smoked link sausage	31.7	389	68
Thuringer	29.9	347	68
Vienna sausage	25.2	279	52

Data based on Composition of Foods, Agricultural Handbook No. 8-7, U.S. Department of Agriculture, Washington, D.C., 1980.

frankfurter, the cholesterol content is twice that of the standard frankfurter due to the high level of mechanically deboned meat which contains high levels of cholesterol.

Similar changes can be obtained by using isolated soy protein to replace fat and meat in a variety of processed meats. The changes which can be obtained with Stolovaya (a Russian large-diameter cooked sausage), pork sausage (an American breakfast sausage), and ground beef (used in hamburgers or meat dishes) are compared in Table 5. In all of the "modified" products, the protein content is increased and the amount of soy protein contributed ranges from 3.4 to 5.5 g per 100-g serving. Fat reductions range from 37 to 52%. Cholesterol content is reduced by amounts ranging from 25 to 28%. The caloric content is reduced as the amount of fat is reduced in the product. These products not only reduce the amount and type of fat which is beneficial for prevention and treatment of heart disease, but reduce the caloric content which is imporant for the prevention and treatment of obesity. The contribution of 3.4 to 5.5 g of soy protein per 100 g of product also has important implications in the treatment of heart disease, since the replacement or addition of soy protein has been documented to lower blood cholesterol levels. This latter observation will be reviewed in more detail in another section of this book.

The amounts of reduction in fat, saturated fat, cholesterol, and calories which can be obtained by consuming two servings of the low-fat, dietetic meat product are given in Table 6. Two servings of 100 g each were used on the assumption that at least two servings per day is not uncommon in developed countries. While the composition of processed meats in a daily menu is infinite, the products in Table 5 have the same general high fat, caloric, and cholesterol content as most of the processed meats given in Table 4. In the U.S., a typical daily menu could include pork sausage for breakfast, bologna for lunch, and ground beef for dinner. The typical U.S.S.R. citizen would consume a frankfurter for breakfast, Stolovaya for lunch, and a sausage for dinner.

The amount of fat reduction from this small and undetectable change in processed foods is similar to the amount of reduction recommended by the World Health Organization, National Institute of Health, and the American Heart Association. The saturated fat is also reduced, but in amounts equal to half that needed to meet the target goal for adult males and 65% of that needed for females. A major reduction in caloric intake is obtained from changing to the low-fat, modified processed meats. This reduction ranges from 86 to 320 kcal per day, depending on the processed meats. The overall effect, if equated to the caloric content of a kilogram of body fat on an annual basis, could have a major effect on obesity. The annual body fat equivalent ranges from 4.1 kg for Stolovaya to 15.5 kg for pork sausage, with a mean of 10.9 kg for the four processed meats.

The use of isolated soy protein, which is virtually fat free, in processed meats to replace some of the high-fat animal products, can provide significant preventive and therapeutic benefits to the consumer. These new protein food ingredients can provide lower fat (leaner meat products) while continuing to maintain the high consumer acceptance without the need to change the cooking and eating habits of consumers.

II. PRODUCTION OF HEALTHIER FOODS WITH ISOLATED SOY PROTEIN

The need for an improved and healthier food supply has been discussed in previous sections. New food products can be formulated with new food proteins to meet the dietary and nutritional requirements of specific segments of the population or the population as a whole. In the previous section, the use of isolated soy protein to modify the composition of processed meats was discussed. In this section, we will discuss how isolated soy protein can be used to modify other food products and how these products can change the composition of dietary intake.

TABLE 5

Nutrient Composition of Traditional and Modified Processed Meats Using Isolated Soy Protein to Replace 25 to 30% of the Traditional Product

Nutrient	Unit	Stolovaya style sausage Amt per 100 g			Pork sausage Amt per 100 g			Ground beef Amt per 100 g		
		Standard	Modified	Percent change	Standard	Modified	Percent change	Standard	Modified	Percent change
Protein	g	13.1	13.6	+4	11.7	17	+45	16.2	19.5	+20
Fat, total	g	21.3	13.5	−37	40.3	20.4	−50	31.0	15.0	−52
Fat, saturated	g	8.1	5.2	−36	14.5	7.3	−50	15.0	7.1	−53
Calories	kcal	249	179	−28	417	254	−39	352	216	−38
Cholesterol	mg	56	40	−28	68	51	−25	68	51	−25
Soy protein	g	0	3.4	—	0	4.4	—	0	5.5	—

TABLE 6

Changes in Intake of Fat, Saturated Fat, Calories, and Cholesterol When Standard Processed Meats Are Replaced by Modified Processed Meats

Dietary component	Unit	Diet and heart target	Frankfurter	Stolovaya	Ground beef	Pork sausages	Average
Fat, total	g	33	−30	−9.6	−32	−39.8	−27.9
Fat, saturated	g	23	−11	−3.6	−16	−14.4	−11.2
Calories	kcal		−254	−86	−272	−326	−234
Cholesterol	mg	200	−44	−26	−34	−34	−34
Soy protein	g		6.8	6.8	11.0	8.8	8.4
Calories annually	kcal		92,700	31,400	99,300	119,000	85,400
Body fat equivalent[a]	kg		12.0	4.1	12.8	15.5	10.9

Note: Reductions based on two servings per day (200 g/d)

a Body fat equivalent calculated based on 7800 kcal per kilogram of fat.

There are several areas that can benefit by the addition of new protein sources to the diet. New food products can be used to meet the recommended dietary changes for prevention and therapy of heart disease in addition to the use of modified meat products. Alternatives to meat and milk proteins can offer even greater benefits in fat and cholesterol reductions than partial replacement obtained with products. In some cases, a combination of two concepts may be the most appropriate for long-term consumption.

High-protein foods to maintain adequate protein intake when low-calorie diets are consumed in weight loss programs are a potential use for new protein isolates. The need for additional protein above the minimum requirement to maintain lean body mass during weight loss has been documented by several research groups.[6-8] The use of modified traditional foods to supply additional protein without adding significantly to calorie intake is an important contribution in the prevention and treatment of obesity and overweight. Obesity has been identified as an additional risk for heart disease, some types of cancers, and contributing to early mortality[9] as discussed in previous sections.

The use of alternate protein sources to replace food proteins to which individuals are allergic is an important food function. Most food proteins have the capability of producing an allergic response if the protein can cross intestinal barriers. Two of the most common allergies are to milk and egg proteins. Intolerance to milk is best documented with infants, although some people are intolerant throughout life. The intolerance to milk can be due to an inability to digest lactose as well as an allergic response to milk protein. In addition, digestive disturbances can result in secondary loss of lactase and lactose intolerance. The use of isolated soy protein as an alternative to milk-based infant formulas is well established in pediatric practice.

New protein foods can also be used in combination with other protein sources to improve the amino acid balance and the protein content of lower-quality proteins. This can be important for specific segments of the population who are underconsuming protein or have an additional need for protein. This can also result in a more efficient use of existing proteins.

A. INFANT FORMULAS

The use of alternate infant formulas to milk has been used for over 50 years. Initially, a number of protein sources were used, including soybean flour for infants who were intolerant of cow milk. Soy flour, due to its high indigestible carboyhdrate content, resulted in digestive disturbances due to fermentation in the lower intestine. In the early 1960s, new infant formulas were produced using isolated soy protein as the protein source with fully digestible carbohydrate from sucrose and/or corn syrup solids (hydrolyzed corn starch). These new formulas improved the acceptability of the products as well as eliminated both the milk protein and the lactose from the formulas. The use and clinical evaluation of isolated soy protein-based infant formulas have been reviewed by Fomon and Ziegler in Chapter 8. The commercial isolated soy protein-based infant formulas have been thoroughly tested in clinical studies and monitored through extensive chemical and microbiological testing to ensure their safety and nutritional adequacy. The use of isolated soy protein-based infant formulas has grown to account for approximately 25% of the infant formulas used in the U.S., based on 1987 to 1989 market surveys.

The composition of a typical isolated soy protein-based infant formula is presented in Table 7. A variety of carbohydrate ingredients as well as fat and mineral sources can be used to meet specific nutrient compositions. The protein content of infant formulas varies between 1.8 and 3.9 g per 100 ml of ready-to-use formulas. The 68 kcal per 100 ml corresponds to the calorie content of human milk. Vitamin and mineral content are set to meet the guidelines of the American Academy of Pediatrics and infant formula regulations. Methionine supplementation has been traditional with isolated soy protein-based infant formulas. However, studies published by Fomon et al.[10] indicate that this is not needed at 2.0 g protein per 100 ml and above.

TABLE 7
Composition of Typical Isolated Soy Protein Infant Formula Powder

Ingredients	g/100 g of dry infant formula powder
Corn syrup solids	30.7
Sucrose	20.5
PP-series protein	16.1
Corn oil	14.0
Coconut oil	14.0
Minerals	2.5
Vitamins and other additives	2.2

Nutrient composition (ready-to-use formula)

	Unit	Amount per 100 ml	Minerals	Unit	Amount per 100 ml
Protein (n × 6.25)	g	1.8	Calcium	mg	70
Moisture	g	90.2	Phosphorus	mg	50
Fat	g	3.7	Potassium	mg	77
Ash	g	0.46	Sodium	mg	32
Carbohydrate	g	6.8	Chloride	mg	59
Calories	kcal	68	Magnesium	mg	5.0
			Iron	mg	1.2
			Zinc	mg	0.5
			Copper	μg	50
			Manganese	μg	20
			Iodine	μg	10

Vitamins	Unit	Amount per 100 ml
Vitamin A	IU	200
Vitamin D	IU	40
Vitamin E	IU	1.7
Vitamin K	μg	10
Vitamin B_1	mg	0.04
Vitamin B_2	mg	0.06
Vitamin B_6	mg	0.04
Vitamin B_{12}	μg	0.3
Vitamin C	mg	5.5
Niacin	mg	1.7
Folic acid	μg	10
Pantothenic acid	mg	0.5
Biotin	μg	3.0
Choline	mg	5.3
Inositol	mg	3.2

Essential amino acid content (ready-to-use formula)

Amino acid	mg per 100 ml	Amino acid	mg per 100 ml
Histidine	47	Cystine	23
Isoleucine	88	Phenylalanine	94
Leucine	148	Tyrosine	68
Lysine	113	Threonine	68
Methionine	24	Tryptophan	25
		Valine	90

B. CEREALS AND PORRIDGES

Cooked breakfast porridge similar to instant cooked wheat or oatmeal has been developed with additional protein, vitamin, and mineral fortification. The taste and acceptance of these products are similar to that of existing products without fortification due to the bland flavor of the isolated soy protein in the formulation. The composition of these products is given in Table 8. The protein content of the high-protein wheat cereal has been increased from the standard 14% protein to 25% protein. Similarly, the high-protein oatmeal has been increased from 16% protein to 23.7% protein. These products have also been fortified with vitamins and minerals to provide 20% of the U.S. recommended daily allowance. The level of vitamin and mineral fortification can, however, be adjusted to meet specific dietary requirements without adversely affecting the acceptability of the food products.

In addition to improving the protein content of the food products, the addition of isolated soy protein to wheat or oats improves the amino acid balance of the cereal. This improves the utilization of the cereal protein, which is limiting in lysine.

In some areas of the world, protein and amino acid inadequacy continues to be a problem due to low income and limited availability of food sources. In many cultures, the populations are primarily dependent on a simple grain for their diet. Grains such as corn and rice are limiting in both total protein and essential amino acid relative to the energy content of these grains. The addition of small amounts of supplemental proteins in the order of 2 to 6% will improve the protein intake to meet minimum requirements and improve the amino acid balance to improve the utilization of the grain protein.

The high-protein cereals can be used as an alternative food for breakfast in place of animal products. The high protein to calorie ratio also makes these products ideal for use in weight-reduction regimes, which require maintenance of high protein intake while reducing calorie intake. Other applications could include supplemental feeding, geriatric feeding, and meal replacement in pre- and postoperative patients where additional protein intake can benefit patients under stress.

Noodle and pasta products can be produced with high protein content (Table 8). The protein content as high as 29% can be produced using standard pasta equipment and production techniques. These products, like the cereals, improve both the protein content and the amino acid balance of the wheat protein used in making pasta. The pasta product in this case was fortified with vitamins and minerals to be equivalent to that of 100 g of meat. Therefore, the pasta or noodles can be used as an alternative to meat in a meal, while providing the nutrition of both the pasta and meat without the addition of fat and cholesterol. A meal of pasta with tomato sauce and vegetables would be an ideal meal replacement for people on weight-loss diets and blood cholesterol-lowering diets. The noodle products could be used either at lunch or dinner meals.

High-protein shapes which are made of 75% isolated soy protein are another way of supplementing meals with concentrated protein sources which are low in fat, calories, and cholesterol. These protein particles can be added to soups, sauces, and casseroles. This can again provide an alternative to meat products or supplement the diets which need additional protein.

Nutritional beverages provide an alternative to milk, which replaces both milk protein, lactose, and milk fat. This type of product is ideal for people who are intolerant to milk or need to reduce the consumption of milk fat and milk protein to reduce blood cholesterol levels. This product can be used at all three main meals and as a supplemental beverage. In addition, it can be used as an alternative to milk in recipes to eliminate both milk protein and lactose from the diet.

Soups made with isolated soy protein can provide a vehicle for supplying protein in a highly acceptable form. A formulation for mushroom soup is given in Table 8. However, a variety of soups can be prepared with different types and flavors. The mushroom soup

TABLE 8A
Composition of Modified Foods Made with Isolated Soy Protein

	High-protein wheat cereal	High-protein oatmeal	High-protein noodles	High-protein pasta	Protein shapes
Supro® protein (g)	18.10	12.00	16.00	14.96	
Whole wheat (g)	79.37	—	—	—	
Wheat flour (g)	—	—	79.00	85.00	
Rolled oats (g)	—	85.88	—	—	
Dry whole eggs (g)	—	—	5.00	—	
Vitamins and minerals (g)	2.53	2.12	—	0.41	—
Total (g)	**100.00**	**100.00**	**100.00**	**100.00**	
Protein (g)	25.0	23.7	29.3	23.9	68.1
Moisture (g)	9.1	8.0	10.4	10.9	10.9
Fat (g)	1.3	6.8	4.1	1.5	3.3
Ash (g)	5.2	3.7	1.6	1.0	3.7
Carbohydrate (g)	59.4	56.6	54.6	63.5	14.0
Calories (kcal)	350	382	372	363	358

TABLE 8B

	Nutritional beverage	Mushroom soup	Cultured desserts
Supro® protein (g)	30.92	39.39	4.40
Nonfat dry milk (g)	—	20.67	—
Instant starch	—	8.27	—
Potato starch (g)	—	6.58	—
Sucrose (g)	39.57	0.94	4.10
Glucose (g)	—	—	1.60
Maltodextrin (g)	—	—	5.60
Fruit base (g)	—	—	18.00
Vegetable oil (g)	15.19	—	1.20
Salt (g)	—	7.52	—
Yeast extract (g)	—	—	0.10
Stabilizer (g)	2.85	0.57	0.10
Vitamins and minerals (g)	11.80	1.02	0.70
Flavor (g)	0.67	15.08	—
Water (g)	—	—	64.20
Total	**100.00**	**100.00**	**100.00**
Protein (g)	26.3	48.5	4.1
Moisture (g)	2.7	2.4	73.2
Fat (g)	18.9	3.5	1.2
Ash (g)	8.3	13.3	1.0
Carbohydrate (g)	39.6	32.3	20.5
Calories (kcal)	419	355	109

presented in Table 8 contains 48.5% protein, while contributing only 3.5% fat on a dry basis. This product would be particularly useful as a protein supplement or alternate protein source. It is, however, a blend of protein sources from both isolated soy protein and nonfat dry milk.

A cultured dessert product has also been developed which can substitute for milk-based desserts, such as ice cream, puddings, and yogurts. The cultured desserts can be fruit flavored and fruit can be added to optimize its taste and acceptance. The product is low in fat and calories, contributing only 10% of the calories from fat and that coming from vegetable

TABLE 9
Daily Menu Developed Using New Food Products Made with Isolated Soy Protein

Menu 1

Breakfast	g	Lunch	g	Dinner	g
Orange juice	250	Bread	50	Tomato sauce	125
Nutritional beverage[a]	250	Stolovaya	100	Pasta[b]	100
Fortified wheat cereal[b]	52	Cucumber	80	Low-fat hamburger	100
Sugar	30	Onion	50	Zucchini	100
Coffee	150	Tomato	125	Broccoli	100
		Soup[b]	26	Bread	25
		Apple juice	250	Tea	250
		Salad dressing	30	Sugar	25
				Margarine	20
				Cultured dessert	170

Menu 2

Breakfast	g	Lunch	g	Dinner	g
Orange juice	250	Bread	50	Soup[b]	26
Nutritional beverage[a]	250	Low-fat frankfurter	100	Low-fat hamburger	100
Fortified oatmeal[b]	80	Green beans	125	Potato	156
Sugar	50	Carrots	80	Cauliflower	100
Coffee	150	Apple	140	Green peas	80
		Coffee	150	Bread	25
		Sugar	20	Cultured dessert	170
		Margarine	10	Margarine	10
				Vegetable oil	10
				Tea	150
				Sugar	20

[a]   Contains 32 g of dry beverage powder and 218 ml of water.
[b]   Dry product before cooking.

oils. The protein content is 4.1 g per 100 g of dessert, which is 15% of the calories. This product can be used at any of the three main meals as an alternative to high-fat, high-calorie desserts.

How well these new food products are able to meet the desired goals of consuming less fat, cholesterol, and calories can best be demonstrated by including them in normal menus. Two typical daily menus are presented in Table 9 using the new food products presented in this section and the modified processed meats presented in the previous section. The foods and amounts of these daily menus are typical of the current consumption of these foods in unmodified foods.

The calorie, protein, fat, fatty acid, and cholesterol content of the meals and daily menus using modified foods are summarized in Table 10. The total calorie content of daily menu 1 is 2353 kcal, 91 mg cholesterol, and 28% of calories from fat. The saturated fat in this product contributes only 7% of total calories.

The protein content of menu 1 is 114.6 g, of which 54.6 g are contributed by isolated soy protein. Daily menu 2 provides 2409 calories, 94 mg of cholesterol, and 28.5% of calories from fat. Saturated fat provides only 7.8% of calories. The menu also provides 107.9 g of protein, of which 41.5 g are from isolated soy protein. The menus provide the fat, calorie, and cholesterol contents needed to meet the recommendations of the World Health Organization, National Institute of Health (U.S.), and the American Heart Association

TABLE 10
Composition of Daily Menus Using New Food Products Made with Isolated Soy Protein

	Calories (kcal)	Protein total (g)	Soy protein (g)	Fat total (g)	Sat. fat (g)	Mono. fat (g)	Poly. fat (g)	P/S ratio	% of cal. from fat	Cholesterol (mg)
Menu 1										
Breakfast	548	23.4	16.75	7.45	1.19	1.81	4.08	3.41	12	0
Lunch	713	32.9	12.5	31.3	6.32	10.8	10.2	1.61	39	40
Dinner	1092	58.3	25.3	34.5	10.53	14.7	9.45	0.90	28	51
Total	**2353**	**114.6**	**54.6**	**73.2**	**18.04**	**27.31**	**23.73**	**1.3**	**28**	**91**
Menu 2										
Breakfast	755	29.4	16.9	13.5	1.92	3.64	6.2	3.23	16	0
Lunch	627	21.7	3.4	25.5	9.09	11.11	4.73	0.52	37	43
Dinner	1027	56.8	21.2	37.3	10.00	13.24	12.90	1.29	33	51
Total	**2409**	**107.9**	**41.5**	**76.3**	**21.01**	**27.99**	**23.83**	**1.13**	**28.5**	**94**

Note: P/S, polyunsaturated fatty acid/saturated fatty acid ratio.

for prevention and dietary treatment of heart disease. In addition, the diet provides significant quantities of soy protein, which has been identified as a dietary aid for lowering cholesterol in a number of clinical evaluations.

REFERENCES

 1. Prevention of Coronary Heart Disease, Report of a WHO Expert Committee, Tech. Rep. Ser. No. 678, World Health Organization, Geneva, 1982.
 2. Health Implications of Obesity, Consensus Development Conference Statement, National Institutes of Health, Vol. 5, No. 9, Washington, D.C., 1986.
 3. Lowering Blood Cholesterol to Prevent Heart Disease, Consensus Development Conference Statement, Vol. 5, No. 7, National Institutes of Health, Washington, D.C., 1985.
 4. Dietary guidelines for healthy adult Americans, *Circulation,* 74, 1465A, 1986.
 5. **Grundy, S. M., Arky, R., Bray, G. A., Brown, W. V., Ernst, N. D., Kwiterovich, P. O., Mattson, K., Weidman, W. H., Schonfeld, G., Strong, J. P., and Weinberger, M.,** Coronary risk factor statement for the American public, a statement of the nutrition committee of the American Heart Association, *Atherosclerosis,* 5, 678A, 1985.
 6. **Bistrian, B. R., Blackburn, G. L., and Stanburg, J. B.,** Metabolic aspects of a protein-sparing modified fact in the dietary management of prader-willi obesity, *N. Engl. J. Med.,* 296, 774, 1977.
 7. **Bistrian, B. R.,** Clinical use of a protein-sparing modified fact, *J. Am. Med. Assoc.,* 240, 2299, 1978.
 8. **Garrow, J. S., Durrant, M., Blaza, S., Royston, D., and Sunkin, S.,** The effect of meal frequency and protein concentration on the composition of weight loss by obese subjects, *Br. J. Nutr.,* 45, 5, 1981.
 9. Health Implications of Obesity, Consensus Development Statement, Vol. 5, No. 9, National Institutes of Health, Bethesda, MD, 1986.
10. **Fomon, S. J., Ziegler, E. E., Nelson, S. E., and Edwards, B. B.,** Requirement for sulfur-containing amino acids in infancy, *J. Nutr.,* 116, 1405, 1986.

Chapter 17

THE EFFECTIVENESS OF LOW-FAT, HIGH-PROTEIN FOOD PRODUCTS IN DIETARY TREATMENT OF OVERWEIGHT AND OBESITY

Michael N. Volgarev, Vadim G. Vysotsky, Taisya A. Yatsyshina, and Valentina A. Meshcheryakova

An analysis of nutrition patterns for populations of various developing and economically advanced countries has shown a relationship between the level of consumption of animal food products containing high saturated fat and cholesterol and the incidence of a number of chronic disorders such as obesity, cardiovascular disease, and diabetes.[1] Data from the American Heart Association (U.S.) indicates that the intake of food exceeds the energy requirement with 40 to 50% of the calorie intake coming from fat, and 15 to 17% of the energy is from saturated fat with a high cholesterol intake of 450 to 500 mg per day.[2] Similar data were observed for selected areas of the U.S.S.R. (the cities of Moscow and Leningrad) during a joint Soviet-American study of population nutrition patterns and the distribution of a number of metabolic and cardiovascular disorders.[3] A direct relationship was found to exist between the consumption of foods with high energy values, primarily due to increases in the amount of animal fats, and the incidence of heart disease, hypertension, and sugar diabetes, which occurred in 40 to 50%, 20%, and 5 to 10% of the population, respectively.

The World Health Organization (WHO) has recommended changes in the consumption patterns for the population, with a view to combating overweight and obesity and preventing certain chronic diseases. These recommendations stipulate a decrease in the caloric values of foods, a decrease in the consumption of saturated fats, and a decrease in dietary cholesterol levels to 300 mg per day.[4,5]

These WHO recommendations may be carried out in two ways: (1) by means of publicizing the principles of proper diet and (2) by producing low-calorie food products with lower cholesterol levels for mass consumption. Under conditions of an industrialized society where population nutrition patterns are more or less based on industrial output of ready-to-cook and prepared food items, the second alternative for prevention and treatment of disorders resulting from excess caloric intake is the basic one to be considered. Once these are available, the publicizing of the proper diet to encourage the use of these products can be implemented.

To meet the above objectives, it is necessary to have lower-calorie protein products containing limited amounts of saturated fat and cholesterol. The most rational means for accomplishing this is by the creation of traditional foods which are fortified with isolated plant proteins. The latter is advantageous, since recent studies have shown that several vegetable proteins, particularly isolated soy proteins, have a marked hypocholesterolemic action as opposed to animal proteins.[6]

The above specifications served as a basis for developing a series of high-protein and low-calorie food items which make up a large part of the diet for the population. These were evaluated as to their effectiveness in clinical and field studies with obese patients and volunteers with weight problems. The five products (Table 1) were prepared with isolated soy protein, 500E (Protein Technologies International), for the study.

Comparison of the chemical composition and caloric value of new and traditional bologna products shows a significant decrease of cholesterol (25%) and animal fat (24%) in the product, a decrease in caloric value (20%), and a significant improvement in protein/fat ratios. A similar tendency was seen when comparing frankfurters, where the new low-fat frankfurter permitted a lower level of cholesterol (26%) and animal fat (50%). The caloric

TABLE 1
Chemical Composition of New and Traditional Food Products

Product		Total protein (%)	Isolated soy protein (%)	Fat (%)	Carbohydrate (%)	kcal/100 g	Cholesterol (mg/100 g)
Bologna-style	Low fat	13.2	5.0	15.0	—	188.0	42.0
sausage	Traditional	13.1	0	19.8	—	231.0	56.0
Frankfurters	Low fat	14.1	5.0	15.0	—	195.0	43.0
	Traditional	12.1	0	30.0	—	322.0	58.0
Oat porridge	High protein	23.7	10.6	6.8	57.4	390.0	—
(dry)	Traditional	16.0	0	6.2	67.0	384.0	—
Pasta (dry)	High protein	23.9	13.3	1.5	69.8	362.0	—
	Traditional	12.0	0	2.0	75.0	366.0	—
Beverage							
Soy drink		3.3[a]	3.3	2.4[b]	5.0[c]	55.0	0
Cow milk		3.3	0	3.3	4.6	61.0	14

[a]   Soy protein.
[b]   Corn oil.
[c]   Sucrose.

value of the new product decreased 40% with significant optimization of the protein/fat ratio. High-protein oatmeal in comparison with the traditional product differs only in increased content of protein of higher biological value with a corresponding decrease in the carbohydrate level. Similar differences are also found in the case of pasta products. The nutritional beverage which was developed does not contain animal protein, fat, or lactose and, in comparison with cow milk, is characterized by a somewhat lower (10%) energy value. In addition, all nonmeat products were enriched with vitamins and minerals.

Products listed in Table 1 served as primary sources of protein and fat in the diets developed for studies with reduced caloric content which also included vegetables, fruits, tea, and coffee. The daily diet included the following products: breakfast — soy drink, high-protein oatmeal, fruit (apples or oranges), and tea or coffee; lunch — meat soup and a high-protein pasta product, bread, low-fat bologna, and coffee or tea; supper — low-fat frankfurters, boiled potatoes, vegetables (cucumbers and lettuce), cheese (fat content 26 to 29%), and coffee or tea. The studies of these diets were conducted under clinical conditions over a period of 30 d on obese patients (clinical diet) and observations with volunteers who carried on a normal way of life (experimental diet).

The proximate[7] chemical composition and caloric value of these diets on a 7-d rotating menu are listed in Table 2. The caloric value of the clinical diet, fed under the controlled conditions of a treatment center, was 25% below that of the experimental diet, mainly due to the fact that the fat content was reduced by half and the carbohydrate level was slightly lower. For the clinical study, 17 male patients were selected who were overweight to the first or second degree. Hospitalization records indicated the following disorders: overweight, exceeding the ideal by 20 to 30%; moderate hypercholesterolemia (6.5 to 7.8 mmol/l); the absence of other disorders (Table 3).

The experimental diet studies were conducted with persons of both sexes (12 women and 2 men) under ordinary conditions of work and leisure, and who expressed a desire to lose weight. The initial characteristics for volunteers are listed in Table 4. The individual deviations in actual body weight from the ideal were wide and included some instances of obesity as well as overweight (Table 5). The majority (71%) of the patients were of first- and second-degree type of obesity and had an absence of overweight individuals, while the volunteers were persons who were overweight or first-degree cases of obesity (71%). The patients also had moderate cholesterolemia as indicated by elevated blood plasma cholesterol,

TABLE 2
Chemical Composition (g/d) and Energy Value (kcal/d) of Low-Calorie Diets

Clinical diet

Days of the week	Protein (g)	Fat (g)	Carbohydrates (g)	Energy (kcal)
1	91.38	40.46	139.91	1289
2	94.12	32.00	135.35	1206
3	96.46	31.85	119.00	1149
4	86.17	38.83	187.16	1442
5	90.22	50.89	153.88	1434
6	96.70	36.73	146.59	1304
7	90.53	40.12	155.85	1346
Mean	92.22	38.69	148.24	1310

Test diet

	Protein (g)	Fat (g)	Carbohydrates (g)	Energy (kcal)
1	90.00	67.80	170.53	1714
2	88.93	72.10	174.79	1762
3	88.42	71.40	154.79	1674
4	88.93	72.10	174.79	1762
5	90.00	67.80	170.53	1714
6	88.93	72.10	174.79	1762
7	90.00	67.80	170.53	1714
Mean	89.32	70.16	170.04	1729

TABLE 3
Characteristics of Obese Patients

No. of subject	Sex	Age (years)	Height (cm)	Body weight (kg)	Percent deviation from ideal	Blood serum cholesterol (mmol/l)	(mg/dl)
				Clinical studies			
1	M	34	180	101.0	36	4.82	186.4
2	M	41	175	109.1	56	8.67	335.2
3	M	46	173	85.8	25	9.27	358.4
4	M	24	180	103.5	39	6.28	242.8
5	M	39	170	114.0	71	6.43	248.6
6	M	33	189	136.5	68	6.88	266.0
7	M	28	176	100.0	41	7.48	289.2
8	M	42	172	98.1	44	5.38	208.0
9	M	31	176	89.7	27	4.90	189.4
10	M	28	179	114.5	56	7.03	271.8
11	M	37	173	99.6	45	6.73	260.2
12	M	31	173	114.5	67	5.61	216.9
13	M	33	180	108.0	46	6.58	254.4
14	M	29	187	98.5	24	6.28	242.8
15	M	43	175	96.5	38	8.37	324.6
16	M	40	178	97.0	34	5.76	222.7
17	M	25	170	92.0	38	5.98	231.2
Mean		34.4	177.0	103.6	44.4	6.61	255.6
SE		+1.5	+1.3	+3.5	+3.3	+0.32	+12.4

TABLE 4
Characteristics of Volunteers

No. of subject	Sex	Age (years)	Height (cm)	Body weight (kg)	Percent deviation from ideal	Blood serum cholesterol (mmol/l)	(mg/dl)
				Volunteer studies			
21	F	56	164	81.5	38	5.20	201.1
22	F	36	160	67.9	21	4.78	184.8
23	F	59	165	71.4	20	5.53	213.8
24	F	46	162	72.2	25	5.46	211.1
25	F	27	167	77.2	27	4.26	164.7
26	F	40	168	89.5	46	3.81	147.3
27	F	31	163	60.0	3	4.94	191.0
28	F	41	171	72.5	14	4.11	158.9
29	F	45	160	92.8	65	4.04	156.2
30	F	55	172	68.0	6	6.88	266.0
31	F	44	160	64.3	14	3.29	127.2
32	F	51	167	82.5	36	5.08	196.4
33	M	23	185	78.5	0	3.59	138.8
34	M	29	170	88.0	30	4.19	162.0
Mean		41.0	167.0	76.2	24.8	4.65	179.8
SE		±3.0	±1.8	±2.5	±4.6	±0.25	±9.7

TABLE 5
The Distribution of Test Subjects by the Type of Body Fat Content

Type of subject	Total number	Normal	Excess body weight	Obese 1st	2nd	3rd
Patients	17	—	—	3	9	5
Volunteers	14	1	4	5	3	1

Note: Excess body weight 100 to 114% of ideal body weight (IBW); first-degree obesity, 115 to 130% IBW; second-degree obesity, 130 to 150% of IBW; third-degree obesity, above 150% of IBW.

averaging 6.61 ± 0.32 mmol/l, while the normal range is 3.9 to 6.3 mmol/l. The volunteers had normal cholesterol values, except for one individual.

All persons participating in a given study were weighed on an empty stomach each day and a doctor conducted a general exam and measured blood pressure, and the overall subjective feelings of the individuals were recorded.

Blood was drawn from patients' arms prior to the experiment and on the 14th and 30th days of the study for biochemical studies. Blood was drawn from volunteers initially and after the 28th day. The blood samples were analyzed for cell count, albumin and triglycerides, transferin, glucose, and the GOT, GPT, and LDH activity with the aid of the standard kits from the firm Boehringer Manheim (West Germany) on an FP-9 analyzer from the firm Labsystem (Finland). Cholesterol was determined by the CHOD-PAD enzymatic colorimetric method using a kit from Sigma (U.S.A.) on a spectrophotometer SF-26 (U.S.S.R.), and also bilirubin, uric acid, Ca, P, Fe, and ALP, γ-GTP activity, and creatine phosphokinase with kits from Yanaco (Japan) on a Raba-Super analyzer of the same firm.*

* GOT, glutamyl oxylate transaminase; GPT, glutamyl pyruvate transaminase; LDH, lactate dehydrogenase; ALP, alkaline phosphatase; γ-GTP, gamma glutamyl transpeptidase.

TABLE 6
Changes in Patient Body Weight

Subject number	Initial		15th day of test		30th day of test	
	kg	% Deviation from ideal	kg	% Deviation from ideal	kg	% Deviation from ideal
1	101.0	36	96.0	29	92.2	24
2	109.1	56	106.0	52	105.0	50
3	85.8	25	84.5	23	80.8	18
4	103.5	39	99.0	33	96.0	29
5	114.0	71	112.2	68	105.5	58
6	136.5	68	132.2	62	123.5	52
7	100.0	41	95.9	35	91.9	30
8	98.1	44	92.6	36	91.0	34
9	89.7	27	86.7	22	85.0	20
10	114.5	56	10.7	51	108.5	48
11	99.6	45	95.5	39	92.9	35
12	114.5	67	112.0	63	110.3	61
13	108.0	46	105.8	43	103.1	39
14	98.5	24	95.5	20	93.0	17
15	96.5	38	90.0	28	88.4	26
16	97.0	34	93.7	26	89.5	
17	92.0	38	89.0	34	86.0	29
Mean	103.6	44.4	99.9	39.1	96.6	34.8
SE	±3.5	±3.3	±3.3	±3.2	±3.0	±3.0
Change from initial values			−3.7	5.3	−7.0	9.6

All patients had their urine and urine creatine tested by the above described blood test methods on days 2 to 4, 6 to 8, 14 to 16, and 28 to 30 of the study. On days 6 to 8 and 28 to 30 of the test, eight patients were also studied for nitrogen balance[9,10] and digestibility[10] of protein. On the 1st, 15th, and 30th days of the study, patients had their shoulder diameter, skin-fold thickness, and shoulder muscle diameter measured, and deviations in these parameters from the norm[8] were calculated. The results of all studies were tested statistically using the Student test.[11]

The clinical observation indicated that the patients tolerated the period of reduced consumption quite well and did not complain of the quality of the products used or of the need for supplementary intake. The values for primary physiological functions for patients (heart rate, blood pressure, gastrointestinal tract functions, and those of the excretory system) did not undergo any changes and were essentially identical to those measured at the outset of testing. One exception was the overall condition of patients which, by their own judgment, improved significantly.

At the conclusion of the study, average weight loss was 7 kg or 233 g per day in the patient group (Table 6). During this period, the percentage amounting to a 20% decline in deviation from ideal body weight declined significantly from the initial values. It should also be emphasized that after the period of diet therapy was completed, there was a shift of the incidence of obesity in the direction of a less serious form. Thus, prior to treatment we had mainly second-degree forms of obesity (nine persons), and the total number of patients with first- and second-degree forms equaled 71%. Upon their discharge from the clinic we had primarily first-degree cases (nine persons), with a total number of patients with second- and third-degree forms of obesity standing at 82%.

Anthropometric measurements conducted during the study also registered specific changes (Table 7). The data given in Table 7 show a gradual decrease in skin-fold thickness and decreased percentage of deviation from the standard value during treatment and an increase in shoulder circumference. However, these parameters were not significantly different.

TABLE 7
Anthrometric Measurement Data

Parameter		Day of test		
		1	15	30
Skin-fold thickness	mm	15.6 ± 1.5	14.3 ± 1.5	13.5 ± 1.4
	Percent deviation from standard	25.2 ± 11.7	14.4 ± 11.7	8.2 ± 11.1
Shoulder circumference	cm	31.0 ± 0.8	31.6 ± 0.6	32.2 ± 0.6
	Percent deviation from standard	˙5.6 ± 2.6	7.8 ± 2.2	9.9 ± 1.9
Shoulder muscle circumference	cm	26.1 ± 0.7	27.1 ± 0.4	27.9 ± 0.5
	Percent deviation from standard	3.2 ± 2.6	7.2 ± 1.7	10.5 ± 2.0[a]

[a]   Significantly different from initial value; $p < 0.05$.

TABLE 8
Biochemical Parameters of Clinic Patients Consuming a 1300-kcal Diet

Biochemical parameters	Day of study			Normal ranges
	Before[a]	14th[a]	30th[a]	
Total protein (g/l)	79.0 ± 1.0	72.0 ± 1.3[b]	72.0 ± 0.6[b]	60—85
Albumin (g/l)	45.0 ± 0.9	42.0 ± 0.6[b]	39.0 ± 0.8[b]	3—43
Transferrin (g/l)	3.10 ± 0.14	2.30 ± 0.11[b]	2.40 ± 0.08[b]	
Urea nitrogen (mmol/l)	3.50 ± 0.23	3.00 ± 0.24	2.40 ± 0.22[b]	4.3—18.6
Creatinine (μmol/l)	102.90 ± 2.95	95.20 ± 3.29	87.80 ± 3.69[b]	44—133
Uric acid (μmol/l)	387.6 ± 18.4	395.1 ± 24.8	353.8 ± 22.0	208—506
Cholesterol (mmol/l)	6.61 ± 0.31	4.27 ± 0.21[b]	3.88 ± 0.19[b]	
Triglycerides (mmol/l)	2.00 ± 0.25	1.48 ± 0.19[b]	1.11 ± 0.19[b]	
Glucose (mmol/l)	5.74 ± 0.20	5.39 ± 0.16	4.40 ± 0.09[b]	3.9—6.4
Bilirubin (μmol/l)	18.80 ± 1.09	—	17.10 ± 1.12	1.7—20.5
Iron (μmol/l)	28.20 ± 1.29	—	25.40 ± 1.66	7.2—26.9
Calcium (mmol/l)	2.35 ± 0.02	—	2.33 ± 0.04	2.1—2.6
Phosphorus (mmol/l)	1.07 ± 0.05	—	1.11 ± 0.04	0.81—1.45
Glutamyl-oxalatetransaminase (nmol/ml)	17.70 ± 1.5	18.82 ± 2.06	17.98 ± 1.45	
Glutamyl-pyruvatetransaminase (nmol/ml)	20.78 ± 2.33	22.06 ± 2.95	23.80 ± 2.52	
GOT/GPT	0.91 ± 0.04	0.96 ± 0.05	0.81 ± 0.04	
Lactate dehydrogenase (nmol/ml)	199.32 ± 7.41	192.22 ± 14.49	165.15 ± 8.11[b]	
Creatine phosphokinase (U/l)	78.00 ± 8.96	125.00 ± 17.27[b]	98.20 ± 15.99	0—225
Gamma-glutamyl transpeptidase (U/l)	30.20 ± 3.75	19.60 ± 2.32[b]	20.70 ± 3.00	0—30
Alkaline phosphatase (U/l)	143.40 ± 7.62	138.50 ± 9.11	165.10 ± 14.20	30—115

[a]   Mean ± standard error of the mean.
[b] $p < 0.05$ in comparison with the first day.

Figures for shoulder muscle circumference showed a tendency to increase slightly, which was linked to a decrease in the amount of subcutaneous fat. Thus, on the basis of anthropometric studies, the principal reasons for body mass loss in obese patients is a decrease in the amount of fat in the body.

In Table 8, the results of biochemical measurements of blood plasma are listed for the patient group. Significant differences did occur in a number of the parameters. We note statistically significant decreases of total protein status based on a decreased level of albumin, urea nitrogen, and creatine. At the same time, some values exceeded the normal physiological limits. This study revealed an incidence of uremia (the normal content of uric acid in the blood is 119 to 238 mmol/l) and normalization by the end of patient testing of the level of transfer in the blood serum.

TABLE 9
Cholesterol Concentrations (mmol/l) in Plasma
Patients

Subject number	Initial	Day 14	Day 30
		Day of test	
1	4.8	2.7	2.8
2	8.7	6.6	5.4
3	9.3	4.5	4.2
4	6.3	5.1	3.7
5	6.4	4.3	3.1
6	6.8	4.5	2.8
7	7.5	4.6	3.3
8	5.4	3.4	2.7
9	4.9	3.7	3.3
10	7.0	3.7	3.3
11	6.7	4.5	4.5
12	5.6	3.5	2.7
13	6.6	3.7	3.7
14	6.3	4.3	4.2
15	8.4	4.6	4.5
16	5.8	4.0	4.2
17	6.0	4.9	3.9
Mean/SE	6.6 ± 0.3	4.3 ± 0.2[a]	3.9 ± 0.2[a]

[a] $p < 0.05$ vs. the initial value.

The measurements of erythrocytes and leukocytes, the hemoglobin content, and the morphological composition of peripheral blood showed no significant deviations. A count of the overall number of lymphocytes (which are one of the criteria for diagnosing protein and protein-energy deficiency) demonstrated that these values remained unchanged during the test period. At the time of patient hospitalization these averaged 1961 ± 175, and at the end of the diet period were 1922 ± 13 lymphocytes for 1 ml^3 of blood.

Thus, data obtained in this study showed that during the diet therapy, certain changes occurred in the biochemical systems of the patients which reflected restructuring of metabolism under conditions of reduced intake of energy and with normal protein intake. However, this restructuring was not profound and quickly assumed an adaptive quality; that is, changes in the parameters studied did not exceed the normal ranges and fluctuations in values for the enzymes involved in protein metabolism and were not changed significantly.

Of all the parameters studied, particular attention should be given the criteria reflecting the status of lipid exchange in the patient's system. At the time patients entered the clinic, most were observed to have moderate hypercholesterolemia, which averaged 6.61 ± 0.3 mmol of cholesterol per liter of blood plasma (the norm is up to 6.3 mmol/l). During diet therapy, the cholesterol level gradually normalized to an average of 3.88 ± 0.19 mmol/l by the end of the study. Thus, total cholesterol decreased by approximately 41% (Table 9). The level of triglycerides in blood plasma also decreased significantly ($p < 0.05$), even though they were within the normal range. This seems to reflect the basic caloric deficiency in the diet. This is also confirmed by a decrease of glucose in the blood over these same periods (Table 8). Values for other parameters for the entire period of the study did not undergo significant changes. Exceptions were lactate dehydrogenase activity, which decreased significantly by the end of the study and went beyond the lower limits of the physiological norms (1800 nmol/l). An unexplained increase at the midpoint occurred in the tests for creatine phosphokinase.

The results obtained point to the presence of a specific hypocholesterolemia activity of

TABLE 10
Nitrogen Metabolism During Weight Loss

	1st period 6 to 8th day	2nd period 28 to 30th day
Nitrogen intake (g/d)	14.7	13.4
Excretion		
Total nitrogen (g/d)		
With urine	12.0 ± 0.5	9.1 ± 0.5[a]
With feces	1.2 ± 0.3	1.2 ± 0.1
True Digestibility[b] (%)	97.6 ± 0.4	97.7 ± 0.6
Nitrogen balance (± g/d)	1.9 ± 0.4	2.8 ± 0.3[a]
Urine urea (mmol/d)	416.6 ± 17.0	302.1 ± 13.7[a]
Urine creatinine (mmol/d)	16.8 ± 0.6	16.3 ± 0.7
Nitrogen utilization (%)	42.1 ± 3.0	53 ± 3.1[a]
Creative growth index	111.7 ± 3.0	108.0 ± 5.5[a]

[a] $p < 0.05$ in comparison to the first day.
[b]   Includes an assumed and constant value of milligrams of nitrogen per kilogram per day for metabolic fecal nitrogen.

the diet, which included several products containing soy proteins. For a final definition of the hypocholesterolemia action of soy proteins, in spite of present published data, it would be advisable to conduct more extensive studies with patients with differing types of hyper-cholesterolemia and with the inclusion of control groups with similar levels of blood cholesterol.

Table 10 gives the data on the study of nitrogen digestibility and nitrogen balance in patients during the initial and final stages of treatment. The balance tests demonstrated a definite deviation in the retention of nitrogen in patients' systems. This was due to the fact that, during the test, apparent nitrogen balance values were measured without regard for nitrogen loss through the skin and by other miscellaneous losses. High temperatures prevailed during the test period (30°C). The positive and somewhat elevated figures for nitrogen balance may be interpreted on the basis of these two causes. Firstly, the high temperature undoubtedly caused an increased loss of nitrogen-containing substances through the skin or other means exclusive of urine and feces. Thus we believe that a slightly excessive rise in positive nitrogen balance is actually less noticeable and is picked up over a period of 6 to 8 d of study of the readings (+1.9 ± 0.4 g/d). This, in principle, is a reflection of the state of nitrogen balance. Secondly, even higher readings for positive nitrogen balance, received from the 28th to the 30th days of the study (+2.8 ± 9.3 g/d), along with the first cause, were possibly linked to the retention of nitrogen in the systems of patients. Confirmation of this supposition may be found in other data gathered, which indicate an increase in utilization of nitrogen and the decrease over the period of uric acid in the blood (Table 7) and the level of its excretion in the urine.

Actual digestibility of proteins from the products being studied turned out to be high and did not differ over the test period (Table 10). Data for the creatine-growth index did not show significant changes in value by the end of the study, and the levels obtained differed only slightly from the ideal. This confirmed the results of anthropometric and biochemical studies relating to the favorable changes in the fat layer.

The reducing effect of the diet on overweight patients was confirmed, and a marked hypocholesterolemic effect was demonstrated during the study. The diet and products maintained health and bodily functions at a practically normal level. This is also supported by the minimal shifts seen in several of the metabolic parameters.

In the studies conducted with volunteers, a significant loss in body weight was also

TABLE 11
Change in Body Weight in Volunteer Group

Subject number	Sex	Initial kg	Initial Percent deviation from ideal	Final kg	Final Percent deviation from ideal	Difference kg	Difference Percent deviation from ideal
21	F	81.5	38	77.3	31	−4.2	7
22	F	67.9	21	64.8	15	−3.1	6
23	F	71.4	20	68.4	15	−3.0	5
24	F	72.2	25	66.6	16	−5.6	9
25	F	77.2	27	74.1	22	−3.1	5
26	F	89.5	46	82.0	34	−7.5	12
27	F	60.0	3	58.8	1	−1.2	2
28	F	72.5	14	67.5	6	−5.0	8
29	F	92.8	65	86.0	53	−6.8	12
30	F	68.0	6	65.0	1	−3.0	5
31	F	64.3	14	62.2	11	−2.1	3
32	F	82.5	36	80.0	32	−2.5	4
33	F	78.5	0	78.0	0	−0.5	0
34	M	88.0	30	79.0	19	−9.0	13
Mean		76.2	24.8	72.1	18.3	−4.1	6.5
SE		±6.6	±4.8	±2.2	±4.0	±0.7	±0.5

observed (Table 11). The data in the table indicate that loss of body weight for volunteers during the test period was 4.1 ± 0.7 kg or 5.4% less than the initial figure. The relative percentage of deviation of actual body weight from the ideal declined a corresponding 26.3%. The marked individuality of this effect should also be noted. In the majority of cases, it resulted in significant loss of body weight for individuals having both great and slight deviations in actual body mass from the ideal.

Biochemical studies of blood serum of volunteers showed definite changes in values for a number of parameters, although the observed deviations at the conclusion of the study as compared to initial values did not exceed normal fluctuations (Table 12). The parameters did not indicate any particular changes, except for a decrease in the concentration of creatine and a slight, but statistically insignificant elevation in the level of uric acid, which was maintained at the upper limit of the physiological norm. As in the clinical tests with patients (Table 9), a significant decrease was observed at the end of the study in the level of cholesterol in blood serum (Table 13). These data lead to the conclusion that the hypocholesterolemic action of soy proteins manifests itself primarily in the presence of hypercholesterolemia and is only slightly in evidence when blood cholesterol levels are normal. It is difficult to explain the recorded changes in activity of glutamyl-oxalate-transaminase and alkaline phosphatase. As opposed to the patients, volunteers did not show deviations in triglyceride and glucose concentrations, which confirms the above stated supposition that the dependence of the levels of such concentrates in blood serum on the amount of fats and carbohydrates consumed, since the amounts of these substances in the diet given to volunteers were higher than that of the patients.

Thus, studies conducted on two groups of people clearly demonstrate the great benefits to be gained from the development and manufacture of a wide assortment of products with high protein content and low caloric values with a view to prevention and treatment of illness related to obesity and excessive body weight. This is supported by the published data and by the experimental results.

TABLE 12
Weight Changes and Biochemical Parameters of Volunteers Consuming a 1700-kcal Diet

Biochemical parameters	Period of study		Normal ranges
	Before therapy[a]	After therapy[a]	
Total protein (g/l)	71.0 ± 2.0	70.0 ± 1.0	60—85
Albumin (g/l)	39.0 ± 1.0	41.0 ± 1.0	33—43
Transferrin (g/l)	2.6 ± 0.1	2.6 ± 0.1	
Urea nitrogen (mmol/l)	2.71 ± 0.11	2.55 ± 0.13	4.3—18.6
Creatinine (μmol/l)	77.5 ± 3.0	56.1 ± 2.7[b]	44—133
Uric acid (μmol/l)	217.2 ± 11.9	224.9 ± 23.8	208—506
Cholesterol (mmol/l)	4.56 ± 0.27	3.91 ± 0.18[b]	
Triglycerides (mmol/l)	0.63 ± 0.03	0.61 ± 0.04	
Glucose (mmol/l)	5.1 ± 0.1	5.0 ± 0.1	3.9—6.4
Bilirubin (μmol/l)	11.6 ± 1.0	13.7 ± 1.2	1.7—20.5
Iron (μmol/l)	18.67 ± 1.47	23.47 ± 2.15	7.2—26.9
Calcium (mmol/l)	2.35 ± 0.05	2.33 ± 0.05	2.1—2.6
Phosphorus (mmol/l)	1.17 ± 0.05	1.20 ± 0.05	0.81—1.45
Glutamyl-oxalate-transaminase (nmol/ml)	7.6 ± 7.7	12.0 ± 0.2[b]	
Glutamyl-pyruvate-transaminase (nmol/ml)	9.8 ± 0.7	8.5 ± 1.3	
GOT/GPT	0.80 ± 0.11	1.50 ± 0.16[b]	
Lactate dehydrogenase (nmol/ml)	158.5 ± 17.0	164.9 ± 5.8	
Creatine phosphokinase (U/l)	36.6 ± 3.5	41.5 ± 3.0	0—225
Gamma-glutamyl transpeptidase (U/l)	11.2 ± 0.8	10.0 ± 1.7	0—30
Alkaline phosphatase (U/l)	108.7 ± 7.1	91.4 ± 7.9[b]	30—115

[a]   Mean ± standard error of the mean.
[b] $p < 0.05$ in comparison with the first day.

TABLE 13
Cholesterol Concentrations (mmol/l) in Plasma of Volunteers

Subject number	Day of test	
	Initial	Day 30
21	5.2	4.5
22	4.8	4.3
23	5.5	4.9
24	5.5	4.2
25	4.3	3.7
26	3.8	3.0
27	4.9	4.6
28	4.1	2.8
29	4.0	3.7
30	6.8	3.5
31	3.3	3.8
32	5.1	5.4
33	3.6	3.8
34	4.2	3.2
Mean ± SE	4.7 ± 0.3	3.9 ± 0.2[a]

[a] $p < 0.05$ vs. the initial value.

REFERENCES

1. Collaborative US-USSR, study of the prevalence of hyperlipoproteinemias and ischemic heart disease in the American and Soviet population, *Am. J. Cardiol.,* 40, 260, 1977.
2. **Ingram, D. D., Thorn, M. D., Stiunett, S. S., and Deev, A. D.,** USSR and U.S. nutrient intake, plasma lipids and lipoproteins in men ages 40—59 sampled from lipid research clinics population forum, in *The Prevention of Cardiovascular Diseases,* Academic Press, New York, 1985.
3. **Klimov, A. N., Gerasimova, Ye. N., Glaszunov, I. S., Deyev, A. D., and Ameteli, M. A.,** *Nutrition and Its Relationship to Cholesterol Lipoproteins of High and Low Density in Selected Sub-Populations in the USSR and the United States,* No. 3, 12-1, 1984.
4. *Energy and Protein Requirements,* Report of a Joint FAO/WHO/UNU Expert Consultation, Tech. Rep. Ser. No. 724, World Health Organization, Geneva, 1985.
5. *Prevention of Coronary Heart Disease,* Report of a WHO Expert Committee, Tech. Rep. Ser. No. 678, World Health Organization, Geneva, 1982.
6. **Carroll, K. K., Huff, M. N., and Roberts, D. C. K.,** Vegetable protein and lipid metabolism, in *Soy Protein and Human Nutrition,* Wilcke, H. L., Hopkins, D. T., and Waggle, D. H., Eds., Academic Press, New York, 1979, 261.
7. *The Chemical Composition of Food Products,* Vol. 1, 2nd ed., Skurikhina, I. M. and Volgarev, M. N., Eds., Agropromizdat, Moscow, 1987.
8. **Blackburn, G., Main, B., and Pierce, E.,** Nutrition in critically ill patient, *Anesthesiology,* 47, 181, 1977.
9. **Ceglarek, M. M., Bryant, B. E., and Whedon, G. A.,** *A Manual for Metabolic Balance Studies,* 1958, 1.
10. **Vysotsky, V. G., Yatsyshina, T. A., Ryamarenkao, T. V., and Mamayeva, Ye. M.,** Methods for determining the biological value of proteins (study), *Medical Reference Journal,* VII, No. 6, 24, 1976.
11. **Kaminsky, L. S.,** The statistical processing of clinical and laboratory data, *Medicine,* Leningrad, 1964.

Chapter 18

DIGESTIBILITY, TOLERANCE, AND METABOLIC EFFICIENCY OF MIXTURED FOOD PRODUCTS MADE WITH ISOLATED SOYA PROTEIN IN HUMANS

Taisya A. Yatsyshina, Oxana A. Plotnikova, and Alexander S. Vitollo

The new sources of food protein being industrially produced in isolated form may be employed for creating blended products with a controlled amino acid composition as closely tailored to the individual needs of humans as possible.[1-3] The improvements in the methodological approaches to choosing more efficient ways of selecting initial components and combinations of these for creating such products confirmed the promise shown by a calculation method based on mutual amino acid enrichment.[4-6] The development of combined protein products takes on particular importance from the standpoint of devising special therapeutic diets, especially antiatherosclerotic diets, using an isolated soya protein known to posses a cholesterol-reducing property.[7-9]

Under controlled conditions at a hospital, men suffering from IIa hyperlipidemia were observed. They each received blended products with an isolated soy protein (500E manufactured by Protein Technologies International, U.S.A.) in their therapeutic diets. The patients, ranging in age from 43 to 56 years, had an average initial weight of 82.0 ± 3.3 kg and height of 170.9 ± 2 cm.

The study lasted for 53 d and was divided into four stages. The first (11 d) was the background period and the three subsequent periods of 14 d each differed in terms of the extent to which the traditional protein products were replaced by those blended with the isolated soya proteins.

In the background period, the patients were put on a clinically developed antiatherosclerotic diet. In the three subsequent periods, two thirds, one third, and one half of the protein (primarily of animal origin) were replaced by blended products containing the 500E isolated soya protein (Table 1).

Three blended products manufactured by Protein Technologies International (U.S.A.) were used in the study: a soya beverage, oatmeal porridge, and noodles. The composition of each product is shown in Table 2.

The added fats in the diets consisted of vegetable oil and butter in a ratio of 1:2, with no more than 20 to 40 g of carbohydrate being employed in the form of sugar. The patients received one "Undevit" multivitamin tablet, dog rose tea, fruits and vegetables, as well as 500 mg of calcium gluconate every day. The patients were under round the clock observation under controlled hospitalized conditions throughout the entire study (53 d), with the usual regimen for daytime and rest periods being followed.

The patients were all given daily medical examinations to ascertain their general conditions, level of satiation, body temperature, arterial pressure, pulse, and respiration rate. The nitrogen balancing evaluations as well as the clinical biochemical examinations were carried out in the last 3 d of each period. The excretion of overall nitrogen and end products of nitrogen metabolism were determined on a daily basis, while the excretion of nitrogen with the feces was determined in composite samples every 3 d. The excretion of endogenous nitrogen in the feces was calculated on the basis of 10.7 mg of nitrogen per 1 kg of body weight per day. The content of nitrogen in the urine, feces, and food rations was determined by the Kjeldahl method. The nitrogen balance and digestibility of the protein were calculated. Generally accepted methods were employed to analyze the excretion of urea, creatinine, and amino nitrogen with the urine.

TABLE 1
Composition and Energy Value of the Diets

Diet component	Control period g	Control period %	Period I g	Period I %	Period II g	Period II %	Period III g	Period III %
Protein (N × 6.25)								
Total	75	100	79.6	100	72.1	100	77.6	100
Test food	0	0	61.2	76.8	24.9	34.5	38.8	47.4
Soya	0	0	23.5	29.5	3.7	5.0	19.9	25.0
Meat	21.0	29	0	0	15.3	21.2	5.0	6.4
Fish	4.7	6.3	0	0	3.6	5.0	4.2	5.4
Other food	49.4	65.9	18.4	23.1	28.3	39.2	31.6	40.7
Lipids	78.0		71.3		75.2		72.2	
Carbohydrates	420		477		425		458	
kcal/d	**2713**		**2897**		**2696**		**2821**	

TABLE 2
Composition and Energy Content of Blended Products
Containing Isolated Soy Protein (500E)

Component	High-protein porridge	High-protein noodles	Soy beverage
Oats	85	—	—
PP500E	15	75	15
Whey	—	—	35
Nonfat dry milk	—	—	50
Wheat flour	—	25	—
Protein	25.0	68.7	35.3
Fat	6.4	0.6	0.8
Carbohydrate	58.0	19.0	52.1
Ash	2.2	3.0	7.5
Crude fiber	1.1	0.2	—
Moisture	9.2	6.7	3.4
Energy content			
Total kcal/100 g	389	356	357
Protein (percent from)	26	77	40
Fats (percent from)	15	2	2
Carbohydrates (percent from)	59	21	58

Some of the patients complained of a bad taste in the mouth and diarrhea during the first week of the period in which two thirds of the protein in the ration was replaced by blended products with isolated soya protein. While these occurrences disappeared as the portion of soya protein in the ration was reduced, one of the patients refused to continue with the diet any longer.

The data concerning the changes in the body weight and indices for nitrogen metabolism are summarized in Table 3. As can be seen from Table 3, there was a tendency for body weight to decrease by period, although the difference was statistically insignificant ($p <$ 0.05). It may be assumed that the reduction in body weight has to do with the inadequacy of energy supply per kilogram of weight, since it peaked (-1.3 kg) in the background period and in the second period when the consumption of energy was the lowest, respectively measuring 34.3 ± 7.4 and 34.6 ± 7.4 kcal per 1 kg of weight as opposed to the recommended level of 46 kcal per kilogram of body weight. The average loss in body weight was lowest (-0.7 kg) in the first period when the supply of energy was higher (36.57 ± 7.85 kcal per kilogram of body weight).

TABLE 3A
Nitrogen Measurement in Type IIa HLP Patients Consuming Blended Protein Products Containing Isolated Soy Protein

Period	Subject	Final body weight (kg)	Nitrogen excreted (g/d)		Endogenous fecal nitrogen	Endogenous urinary nitrogen
			Urine	Feces		
			Control			
12 g N per day	1	73.6	9.81	1.04	0.79	2.63
	2	70.0	9.16	1.37	0.59	2.50
	3	91.1	10.81	1.09	0.97	3.26
	4	87.6	9.30	2.17	0.94	3.14
	5	79.9	10.19	1.55	0.85	2.86
	6	84.0	10.33	1.21	0.90	3.00
Mean ± SE		81.0 ± 0.26	9.93 ± 0.17	1.41 ± 0.17		
			I			
12.73 g per day	1	70.6	9.86	1.03	0.76	2.53
	2	69.3	8.26	1.41	0.74	2.48
	3	89.1	9.14	1.29	0.96	3.19
	4	86.9	11.52	2.38	0.93	3.11
	5	79.0	10.19	1.57	0.85	2.83
	6	85.0	6.08	1.73	0.91	3.04
Mean ± SE		81.0 ± 3.5	9.93 ± 0.76	1.41 ± 0.19		

TABLE 3B
Nitrogen Metabolism Indices in Type IIa HLP Patients Consuming Blended Protein Products Containing Isolated Soy Protein

Period	Nitrogen balance (g/d)	Digestibility (%)	Net protein utilization (%)	Biological value (%)
	Control			
1	1.15	97.9	38.1	38.9
2	1.47	94.8	39.3	41.5
3	0.10	99.0	36.1	36.4
4	0.53	89.8	38.4	42.8
5	0.26	94.2	33.1	35.1
6	0.46	97.4	36.3	37.3
Mean ± SE	0.66 ± 0.22	95.5 ± 1.37	36.9 ± 0.9	38.7 ± 1.2
	I			
1	1.84	97.9	40.3	41.2
2	1.84	97.9	40.3	52.1
3	2.30	97.4	50.7	52.0
4	−1.17	88.6	22.6	25.4
5	0.97	94.3	36.5	38.7
6	4.92	93.6	69.7	74.6
Mean ± SE	1.99 ± 0.83	94.4 ± 1.36	44.9 ± 6.47	47.3 ± 6.8

The results obtained from an examination of nitrogen excretion in the feces (Table 3) showed the value for actual true digestibility to be quite high (95.5 to 95.9%) in the background and second group when the level of protein consumption was the lowest (12 and 11.53 g/d based on nitrogen, respectively) and decreased somewhat (93.4 to 94.4%) in the first and third periods when the content of protein was higher (12.73 to 12.4 g/d). As

TABLE 4
End Products of Nitrogen Metabolism at Various Levels of Isolated Soya Protein (PP500E)

Nitrogen metabolite		Control period	Period I	Period II	Period III
Total nitrogen	g/d	9.92 ± 0.26	9.18 ± 0.76	9.49 ± 0.61	9.38 ± 0.33
Urea	g/d	18.63 ± 0.80	17.50 ± 0.90	18.15 ± 1.10	17.83 ± 1.20
Urinary N	g/d	8.69 ± 0.38	8.17 ± 0.43	8.50 ± 0.50	8.32 ± 0.55
Percent of total N	%	88.80 ± 3.00	88.50 ± 3.90	89.80 ± 1.40	88.33 ± 1.93
Creatinine	g/d	1.19 ± 0.06	1.15 ± 0.08	0.99 ± 0.12	1.15 ± 0.03
Creatinine N	g/d	0.44 ± 0.02	0.43 ± 0.03	0.37 ± 0.05	0.42 ± 0.01
Percent of total N	%	4.43 ± 0.23	4.63 ± 0.34	3.86 ± 0.48	4.51 ± 0.11
Creatine	g/d	0.22 ± 0.05	0.15 ± 0.05	0.13 ± 0.04	0.15 ± 0.09
Amine N	g/d	0.29 ± 0.01	0.25 ± 0.02	0.27 ± 0.01	0.28 ± 0.01
Amino O	g/d	0.25 ± 0.01	0.22 ± 0.02	0.24 ± 0.01	0.24 ± 0.01
Percent of total N	%	2.52 ± 0.08	2.45 ± 0.21	2.53 ± 0.13	2.61 ± 0.14

had been assumed, the value for true digestibility was found on average to be within the limits typical for proteins of milk-vegetable origin.

The excretion of total nitrogen in the urine during the background period was somewhat higher (9.93 g/d) than in the subsequent periods when blended products high in protein content were included in the rations (9.18, 9.49, 9.38 g/d). It is interesting to note that the excretion of urea in the urine remained uniform by period, and the proportion of nitrogen in the urea from the total nitrogen comprised 88.8, 88.5, and 88.3%, respectively, in the background, period I, and period III, and was the highest (89.8%) in the second period when the level of protein consumption was the lowest at 11.53 g/d (Table 4). The level of creatinine excretion was the lowest in this period (0.37 g/d, making up 3.86% of the total nitrogen as opposed to 4.43, 4.63, and 4.51% for the remaining periods). Of interest here is the nitrogen balance index (Table 3), which increased significantly in the first and third periods of the examination when the level of protein in the ration and the proportion of animal proteins replaced by blended products high in protein was the highest. Having more than tripled in the first period in comparison to the background, the retention of nitrogen at a comparatively low level of nitrogen consumption may apparently be explained by the sufficient balance of the products high in protein in terms of the amino acid composition, in particular by the high level of sulfur-containing amino acids (porridge: 3.5 g per 100 g of protein; beverage: 3.2 g per 100 g of protein; noodle soup [Shapes]: 2.7 g per 100 g of protein).

This hypothesis is also corroborated by the results obtained from biochemical examinations of the blood serum, which made it possible to trace the statistically significant ($p < 0.05$) increase in the level of serum protein and albumin (Table 5). Of note as well was the increase in the concentration of uric acid in the blood serum, which rose from 3.99 to 5.74 mg/dl.

Special attention should be focused on areas which might explain the possible mechanisms by which changes occur in the level of total cholesterol and neutral lipids as well as of alpha and pre-beta-cholesterol. The reduction in the content of total cholesterol and neutral fats peaked in the first and third groups where the proportion of protein replaced by products containing isolated soy protein was the highest. Some increase in these indices was observed to occur in the second period. It was precisely during this time in which the proportion of soya protein was significantly lower in comparison to the level in the first and third periods.

The fluctuating levels of the various indices for lipid metabolism could not be attributed solely to the shortage of energy, since the maximum increase in the content of total cholesterol

TABLE 5

Biochemical Measurements of Blood Serum at Various Levels of Isolated Soy Protein Intake

Biochemical assay	Units	Normal range	Control period	Period I	Period II	Period III
Total protein	g/dl	6.5—8.0	6.51 ± 0.14	7.02 ± 0.13	7.45 ± 0.13	7.58 ± 0.11
Albumin	g/dl	3.7—5.2	4.27 ± 0.12	4.28 ± 0.06	4.80 ± 0.07	4.65 ± 0.05
Glucose	mg/dl	60—100	86.15 ± 2.43	85.46 ± 1.83	90.37 ± 2.59	91.70 ± 1.46
Bilirubin, total	mg/dl	0.3—1.1	0.64 ± 0.07	0.66 ± 0.05	0.63 ± 0.22	0.68 ± 0.06
Total Urea N	mg/dl	8—20	15.01 ± 0.65	14.92 ± 0.63	16.10 ± 0.76	15.92 ± 0.72
Creatinine	mg/dl	0.8—1.5	1.45 ± 0.03	1.24 ± 0.02	1.32 ± 0.03	1.19 ± 0.03
Urea N:creatinine ratio			10.42 ± 0.51	12.16 ± 0.84	12.34 ± 0.68	13.56 ± 0.74
Creatine kinase	U/l	5—80	66.14 ± 7.25	54.07 ± 7.68	53.01 ± 7.79	50.07 ± 7.33
Aspartate transaminase	Carmen units	8—35	27.71 ± 0.97	27.69 ± 0.93	36.35 ± 1.54	32.61 ± 1.91
Alanine transaminase	Carmen units	5—30	19.86 ± 2.41	20.51 ± 1.43	21.85 ± 1.79	22.19 ± 3.05
Uric acid	mg/dl	3.7—6.9	3.99 ± 0.19	5.22 ± 0.17	5.74 ± 0.18	5.50 ± 0.16
Cholesterol, total	mg/dl	150—250	313.63 ± 14.30	267.18 ± 7.54	269.96 ± 3.97	237.41 ± 4.37
Triglycerides	mg/dl	40—120	154.10 ± 11.40	159.33 ± 17.70	187.93 ± 18.20	170.82 ± 20.20
Calcium	mg/dl	8.8—10.2	9.10 ± 0.36	10.05 ± 0.15	10.35 ± 0.22	10.91 ± 0.30
Iron	mg/dl	70—200	142.80 ± 5.01	163.26 ± 6.81	123.86 ± 7.29	160.97 ± 8.67

and neutral fats in the blood serum drops in the second period when the loss in body weight is the greatest (-1.3 kg).

The reduction in the level of total blood cholesterol measured 11% in both the first and third periods of the observation, while no decrease in hypercholesterolemia occurred in the background and second periods, but a tendency toward an increase in all lipoprotein fractions in the blood took place. The content dropped 29% (in the first period) and 21% (in the third period) in very low-density lipoproteins (pre-LDL cholesterol), while these levels were 12 and 10% for low-density lipoproteins (LDL-cholesterol) in the first and third periods of observation, comprising 18 and 8% of the initial level. For this reason, the ratio of LDL-cholesterol to total cholesterol determining the activity of atherogenesis remained virtually unchanged during each period of the observation. The direction and extent of the changes which took place in the level of triglycerides were consistent with the dynamics of cholesterolemia.

The mechanism underlying the hypolipidemic effect exerted by isolated soya protein is apparently very involved and requires further study. But we may assume for the present that the selective capacity of isolated soya protein to influence the composition of apoproteins lies at the heart of this mechanism.[9,10] Mixtures of amino acids from proteins of vegetable or animal origin are known to impact the level of cholesterol in the blood in different ways.[7,11] At the same time, intact isolated soya proteins are superior to amino acid mixtures in terms of the hypocholesterolemic effect. The amino acid composition of soya protein is apparently one of the important factors impacting a hypocholesterolemic effect, but not the only one.[7,9]

REFERENCES

1. **Goldberg, A. P., Lim, A., Kolar, J. B., Grundhauser, J. J., Steinke, F. H., and Schonfeld, G. et al.,** Soybean protein independently lowers plasma cholesterol levels in primary hypercholesterolemic, *Atherosclerosis,* 43, 355, 1982.
2. **Scrimshaw, N. S. and Young, V. R.,** Soy protein in adult human nutrition: a review with new data, in *Soy Protein and Human Nutrition,* Wilcke, H. L., Hopkins, D. T., and Waggle, D. H., Eds., Academic Press, New York, 1979, 121.
3. **Sirtori, C. R., Gatti, E., Mantero, O., Conti, F., Agradi, B. S. et al.,** Clinical experience with the soybean protein diet in the treatment of hypercholesterolemia, *Am. J. Clin. Nutr.,* 32, 1645, 1979.
4. **Sugano, M., Tanaka, K., and Ide, T. J.,** Secretion of cholesterol, triglyceride and apolipoprotein A-I by isolated perfused liver from rats fed soybean protein and casein or their amino acid mixtures, *J. Nutr.,* 112, 855, 1982.
5. **Yatsyshina, T. A., Kalamkarova, O. M., and Ivashchenko, N. V.,** Effect of food protein on blood serum cholesterol level (review), *Med. Ref. Zh.,* I, No. 7, 35, 1985.
6. **Tolstoguzov, V. B.,** Synthetic food products, *Science,* Moscow, 9, 1978.
7. **Shaternikov, V. A.,** Medical-biological aspects of the problem of enriched food protein, in *Theoretical and Clinical Aspects of Science with Respect to Nutrition,* Shaternikov, V. A., Ed., Moscow, 1980, 134.
8. **Shaternikov, V. A., Vysotskii, V. G., Yatsyshina, T. A. et al.,** Ways to increase biological value of vegetable origin proteins, *Vopr. Pitan.,* 7, 35, 1982.
9. Foods and Nutrition: A Strategy Within the Framework of National Development, Ninth Report of the Joint Committee of Experts of the FAO/WHO on Nutrition, Tech. Rep. Ser. No. 584, World Health Organization, Geneva, 1977.
10. **Wolf, B. M., Giovannetti, P. M., Cheng, D. C. H., Roberts, D. C. K., and Carroll, K. K.,** Hypolipidemic effect of substituting protein isolate for all meat and dairy protein in diets of hypercholesterolemic men, *Nutr. Rep. Int.,* 246, 1187, 1981.
11. **Pokrovsky, A. A.,** Biochemical backgrounds for the development of products possessing high biological value, *Vopr. Pitan.,* 1, 3, 1964.

Chapter 19

CLINICAL AND EXPERIMENTAL APPROACHES TO STUDYING THE HYPOCHOLESTEROLEMIC EFFECTS OF SOYBEAN PROTEINS

Boris G. Lyapkov

Studies carried out by numerous authors demonstrated that diets which include vegetable proteins produce a reduction in the level of cholesterol in the blood plasma as compared to diets containing casein.[1,2] Of the vegetable proteins, soya proteins exert the most pronounced cholesterol-reducing effect.[3,4] Experiments on rabbits showed that a low level of cholesterol in the blood plasma is brought about by the reduced absorption of cholesterol in the intestine and the increased removal of sterols from the circulatory system by the liver, both of which are caused by an increase in the content of beta-cytosterol and fiber as well as by the predomination of unsaturated fatty acids in the fatty component.[5-7]A hypocholesterolemic effect was also discovered when employing a mixture of amino acids identical in composition to the soya protein isolate in the ration.[8] It was suggested with respect to this datum[9,10] that the hypocholesterolemic effect exerted by soya proteins has to do with the distinguishing features of their amino acid composition, in particular with the low value of the lysine/arginine ratio. Virtually no studies have been conducted to date on the interrelation and mechanisms of the changes which take place in protein and lipid metabolism when significant amounts of soya proteins are included in the diet.

This study examined the effect of diets formulated with isolated soya protein on the content of free amino acids, several protein metabolites in the blood plasma and urine, the level of cholesterol in various fractions of lipoprotein in the blood plasma, and the rate of synthesis of the very low-density lipoproteins (VLDL) and the high-density lipoproteins (HDL).

While under clinical observation at a hospital, 16 patients with unstable ischemic heart disease and type IIa hyperlipidemia (beta-lipoprotein fraction predominated in polyacrylamide gel (PAG) during electrophoresis), aged 43 to 56 years (average initial body weight 82.0 ± 3.3 kg, height 170.9 ± 2.0 cm), were put on an antiatherosclerotic diet of natural products containing 70 to 75 g of protein at an average energy value of 2782 kcal (background period) for a period of 11 d. Over the next 14 d, two thirds of the protein in the diet (primarily of animal origin) was replaced by a combination of protein products containing an isolated soya protein (500E, Protein Technologies International, U.S.A.) which contributed 29.5% of the dietary protein (period of observation). The content of free amino acids in the blood plasma and urine, as well as several metabolites from protein metabolism, was determined at the end of each dietary period. The overall level of cholesterol and neutral fats in the blood plasma was also measured. The amino acid composition of the blood plasma and urine was analyzed and the metabolites from protein metabolism were determined according to a physiological program (44 identified compounds) with a model 6300 amino acid analyzer manufactured by the Beckman Company using the reagents, standards, and columns of the latter. The samples for analysis were prepared in accordance with the recommendations of the Beckman Company. The plasma proteins were precipitated with 50% sulfosalicylic acid, after which the precipitate was removed by centrifugation and the supernatant acidified with a buffer solution for diluting the sample (pH 2.2). The urine was filtered through an antimicrobial filter and brought to a constant volume, from which 1 ml was taken and processed with sulfosalicylic acid (30% solution, 10:1 by volume). Following centrifugation, the supernatant was acidified with a buffer solution (pH 2.2). The chromatograms were measured with a Chromatopak C-RIB® integrator, made by the Shimadza company, using *S*-2-amino-ethyl-L-cysteine as the internal standard. A Technicon® analyzer was employed to determine the overall level of cholesterol and neutral fats in the blood plasma.

The rate of VLDL apoprotein synthesis was determined in experiments with male Wistar rats. The rats were fed nutritionally complete, semisynthetic rations containing the proteins (at 18%) of total calories, from either casein or isolated soy protein) for 30 d *ad libitum*. The rate at which radioactive methionine was included in VLDL apoproteins after exposure 1 h after intraperitoneal injection of the isotope and the half-life of HDL apoproteins in the blood plasma was determined after introduction of radioactive amino acid at 1 and 3 d. The lipoproteins of the blood plasma were isolated in a density gradient of saline solutions during the process of ultracentrifugation according to the method of Hinton et al.[11] The lipids were extracted from the fractions obtained with a butanoldiethyl ether mixture (4:6), after which the extract was evaporated and the precipitate dissolved in 0.5 ml of chloroform. The cholesterol was determined with the Libermann-Burkhardt reaction. The proteins in the samples were precipitated with 5% trichloroacetic acid and removed by centrifugation, after which the precipitate was rinsed twice with an ethanol-diethyl ether mixture (1:1) and dissolved in 0.05 N KOH. The protein content was determined by the Louri method. The radioactivity of the samples was measured in a scintillation medium protein solubilizer (PCS) (Nuclear-Chicago® Solubilizer, Amersham, England) using a Rakbeta counter manufactured by the LKB-Ballak Company. The actual radioactivity was calculated by taking into account the quenching of scintillation determined via the "Hat-trick" method according to the quenching of internal standards (^{3}H and ^{14}C) by carbon tetrachloride. The half-life was calculated by the method of Shimke[12] using data from simultaneous recordings of tritium and carbon ^{14}C in apoproteins of the isolated apoproteins. We described the method in some detail earlier.[13]

Data concerning the content of free amino acids and several metabolites from protein metabolism in the blood plasma and urine of patients in both periods of the examination are presented in Table 1. The overall quantity of essential amino acids in the blood plasma over the period of observation did not differ significantly from the background quantity, while the concentration of lysine and arginine was lower. In addition, there was an increase in the concentration of isoleucine, phenylalanine, and tyrosine. No major changes were observed in the level of the sulfur-containing amino acids methionine and cystine in the blood plasma. The most informative with respect to the nitrogen balance, the valine/arginine index increased somewhat over the period of observation, which was in agreement with the data concerning the direct determination of the nitrogen balance. The concentration of uric acid in the blood rose, while that of creatinine dropped. In this period a pronounced reduction in the concentration of total cholesterol in the blood plasma (2.25 + 0.8 g/l in the background period, 2.01 + 0.08 g/l in the period of observation) was observed, but no change in the content of neutral fats (background: 1.54 ± 0.11 g/l; period of observation: 1.59 ± 0.17 g/l).

While examining the excretion of amino acids from the urine, a reduction in the elimination of two essential amino acids, valine and leucine, was found; the excretion of glycine and cystine from the urine was lower than the control level. The excretion of 3-methyl histidine from the urine did not change significantly over the period of observation, which indicates that there were no changes in the rate of catabolism in muscle tissue.

It was discovered during experiments with rats given a ration with soya protein that the rate at which ^{14}C methionine is included in VLDL apoproteins in the blood plasma is lower than when using a ration with casein (Table 2). The half-life period of HDL apoproteins was high in the experimental group of animals, which points to a deceleration in the renewal rate of this lipoprotein fraction in the blood plasma. The content of cholesterol in the HDL of rats in the experimental group was no different than that of the control group, while a

TABLE 1

Content of Free Amino Acids in the Blood and Urine of Patients (Mean ± SE)

Index in question	Blood (n = 16)		Urine (n = 6)	
	Background period 2	Period of observation	Background period	Period of observation
1	2	3	4	5
Valine	0.271 ± 0.018	0.261 ± 0.027	131.9 ± 24.1	69.9[a] ± 12.9
Leucine	0.231 ± 0.016	0.264 ± 0.008	169.8 ± 29.7	100.7[a] ± 13.7
Isoleucine	0.089 ± 0.005	0.145[a] ± 0.007	—	—
Methionine	0.024 ± 0.002	0.030 ± 0.003	69.6 ± 18.3	83.4 ± 11.4
Cystine	0.050 ± 0.007	0.036 ± 0.001	29.2 ± 5.5	21.5[a] ± 2.4
Threonine	0.132 ± 0.006	0.143 ± 0.009	111.0 ± 14.8	118.4 ± 23.1
Phenylalanine	0.063 ± 0.003	0.080[a] ± 0.006	102.1 ± 17.7	69.4 ± 18.1
Tyrosine	0.066 ± 0.004	0.080 ± 0.004	142.6 ± 19.6	130.1 ± 22.1
Lysine	0.173 ± 0.010	0.147[a] ± 0.005	86.2 ± 13.1	53.2 ± 18.3
Tryptophan glycine	0.287 ± 0.024	0.292 ± 0.027	1660.5 ± 162.8	1013.6[a] ± 259.9
Alanine	0.432 ± 0.044	0.450 ± 0.051	299.1 ± 69.3	258.4 ± 49.2
Arginine	0.127 ± 0.010	0.072 ± 0.040	—	—
Histidine	0.082 ± 0.003	0.093 ± 0.008	—	—
Serine	0.158 ± 0.010	0.141 ± 0.005	295.5 ± 61.0	274.7 ± 75.1
Proline	0.202 ± 0.026	0.220 ± 0.018	—	—
Aspartic acid	0.202 ± 0.026	0.220 ± 0.018	—	—
Ornithine	0.083 ± 0.006	0.087 ± 0.004	—	—
Citruline	0.017 ± 0.003	0.037[a] ± 0.008	—	—
Urea	5.382 ± 0.290	4.706 ± 0.294		
3-Methyl histidine			324.6 ± 136.7	279.5 ± 64.2
Anserine			124.5 ± 19.5	34.6[a] ± 23.5
Uric acid	0.040 ± 0.002	0.052[a] ± 0.002		
Creatinine	0.015 ± 0.003	0.012[a] ± 0.002		
Total essential amino acids	1.097 ± 0.041	1.168 ± 0.030		
Valine/arginine index	2.280 ± 0.200	3.940[a] ± 0.400		
Lysine/arginine index	1.510 ± 0.150	2.140[a] ± 0.130		

Note: The content of amino acids is given in micromoles per milliliter for the blood and in micromoles per day in the urine, while the content of uric acid and creatinine are expressed in grams per liter.

[a] Significant difference, $p < 0.05$.

tendency for the cholesterol level to drop was observed in VLDL and low-density lipoproteins (LDL).

One may conclude from the data obtained that using a significant quantity of soya proteins in a ration results in a change of amino acids in the blood plasma manifested by a reduction in the concentration of lysine and arginine and a drop in isoleucine, phenylalanine, and tyrosine as well as a change in the excretion of individual amino acids from the urine. This study also showed a positive nitrogen balance and the absence of any pronounced changes in the intensity with which contractile proteins of the muscle tissue undergo catabolism. The imbalance of amino acids indicated probably has to do with the varying ease of access to proteolysis on the part of individual amino acids in an isolated soya protein (lysine and arginine in particular), the distinguishing features of the amino acid composition of the soya protein, the differing intensity with which tissue proteins undergo catabolism, and the reuse of endogenously formed amino acids making up the majority of the amino acid pool in the blood plasma.

Is there a connection between the changes in amino acid metabolism indicated above and the hypocholesterolemic effect of the isolated soya protein which surfaced during the

TABLE 2
Content of Overall Cholesterol in the Lipid Proteins of the Blood Plasma in Rats,
Specific Radioactivity of Lipoprotein Apoproteins and their Half-Life Period

Index	VLDL		LDL		HDL	
	Control	Experiment	Control	Experiment	Control	Experiment
1	2	3	4	5	6	7
Total cholesterol (g/l) (n = 12)	0.046 ± 0.006	0.035 ± 0.004	0.120 ± 0.010	0.090 ± 0.008	0.43 ± 0.05	0.40 ± 0.05
Specific radioactivity of lipoprotein apoproteins (total), imp/min[c] per 1 mg protein	$11,705 \pm 981$	$8,488^a \pm 320$			5,065—1,570[b]	5,983—2,671[b]
Half-life period of apoproteins (total), in days^{-1}					1.77[d]	2.36[d]

Note: VLDL, very low density lipoprotein; LDL, low density lipoprotein; HDL, high density lipoprotein.

[a] $p < 0.05$.
[b] First number: specific radioactivity after 1 d, second number: 3 d after the introduction of U^3H and ^{14}C methionine.
[c] imp/min, impulse per minute.
[d] Average of three determinations.

study? It would seem possible to assume that the reduction in the concentration of lysine and arginine in the plasma pool can limit the synthesis in the liver of apoprotein E, a protein rich in arginine,[14] which is a receptor of tissue cholesterol[15] and participates directly in its transportation in the bloodstream. The apoprotein E belongs to both the HDL and VLDL groups. Characteristically, we discovered a change in the parameters of both fractions while studying the metabolic rate of apoproteins VLDL and HDL in rats given a ration containing isolated soya protein. This indirectly corroborates our hypothesis and indicates that the hypocholesterolemic effect exerted by isolated soya protein comes about due to a change in the metabolism of both atherogenic as well as antiatherogenic classes of lipoproteins. In our experiments with rats we observed a reduction in the content of cholesterol in the VLDL and HDL of blood plasma. Apparently, other factors including the ones mentioned earlier enter into the mechanism behind the hypocholesterolemic effect exerted by soy proteins.[5-7]

Hence, diets with soya proteins trigger an amino acid change in the blood plasma which manifests itself in a reduction in the concentration of lysine and arginine. At the same time, a drop in the production rate of VLDL apoproteins was observed and HDL apoproteins are replenished in the blood plasma. It is postulated that the insufficiency of lysine and arginine under the conditions indicated brings about a reduction in the synthesis and metabolism of apoprotein E, which plays an important part in the cholesterol receptor role of HDL. This in turn leads to a decrease in the concentration of blood cholesterol.

REFERENCES

1. **Carroll, K. K.,** The role of dietary protein in hypercholesterolemia and atherosclerosis, *Lipids,* 13, 360, 1978.
2. **Kritchevsky, D., Tepper, S. A., Czarnecki, S. K., Klurfeld, D. M., and Story, J. A.,** Experimental atherosclerosis in rabbits fed cholesterol-free diets; beef protein and textured vegetable protein, *Atherosclerosis,* 39, 169, 1981.
3. **Carroll, K. K., Givannetti, P. M., Huff, M. W. et al.,** Hypercholesterolemia effect of substituting soybean protein for animal protein in the diet of healthy young women, *Am. J. Clin. Nutr.,* 31, 1312, 1978.
4. **Fumagalli, R., Paoletti, R., and Howard, A. N.,** Hypercholesterolemia effect of soya, *Life Sci.,* 22, 947, 1978.
5. **Huff, H. W. and Carroll, K. K.,** Effects of dietary proteins and amino acids mixtures on plasma cholesterol levels in rabbits, *J. Nutr.,* 110, 1676, 1980.
6. **Pathirana, C., Gibney, M. J., and Taylor, T. G.,** Effects of soy protein and saponins on serum and liver cholesterol in rats, *Atherosclerosis,* 36, 595, 1980.
7. **Sklan, D., Budowski, P., and Hurwitz, S.,** Absorption of oleic and taurocholic acids from the intestine of the chick, *Biochim. Biophys. Acta,* 573, 31, 1979.
8. **Jackson, R. L., Morrisett, S. D., and Gotto, A. M.,** Lipoprotein structure and metabolism, *Physiol. Rev.,* 56, 259, 1976.
9. **Huff, H. W. and Carroll, K. K.,** Effects of dietary protein on turnover oxidation and absorption of cholesterol and on steroid excretion in rabbits, *J. Lipid Res.,* 21, 546, 1980.
10. **Sugano, M., Tanoka, K., and Ide, T.,** Secretion of cholesterol, triglyceride and apolipoprotein A-1 by isolated perfused liver from rats fed soybean protein and casein or their amino acid mixtures, *J. Nutr.,* 112, 855, 1982.
11. **Hinton, R. H., Al-Tamer, Y., Mallinson, A., and Marks, V.,** The use of density gradient centrifugation for the separation of serum lipoproteins, *Clin. Chim. Acta,* 44, 267, 1973.
12. **Schimke, R. T.,** Turnover of membrane proteins in animal cells, in *Methods in Membrane Biology,* Vol. 3, Schultz, J. and Block, R. E., Eds., Plenum Press, New York, 1975, 201.
13. **Lyapkov, B. G.,** The rate of apoprotein metabolism of high density lipoproteins from blood plasma of rats in normal state and after treatment with the alimentary factor, *Vop. Med. Chem.,* 5, 686, 1981.
14. **Kritchevsky, D. et al.,** Experimental atherosclerosis in rabbits fed cholesterol-free diets, *Atherosclerosis,* 26, 397, 1977.
15. **Titova, G. V., Kivueva, N. N., and Klimov, A. N.,** Interaction of cholesterol with apoprotein E — an arginine-rich protein of very low density lipoproteins, *Biochimia,* 45, 51, 1980.

Chapter 20

REVIEW OF THE STUDIES OF SOY PROTEINS FOR REDUCING SERUM CHOLESTEROL

Fred H. Steinke

The recognition that diet can play an important role in the prevention and treatment of heart disease has been documented for a number of years.[1] Many of the early studies comparing epidemiological evaluations of populations, countries, and dietary factors led to the conclusion that heart disease was related to serum cholesterol levels and that serum cholesterol levels could be related to dietary factors as well as specific genetic defects.[1,2] The Japanese living in Japan, when compared to similar individuals of Japanese ancestry living elsewhere in the world, show markedly different serum cholesterol levels and different incidences of heart disease, which gives clear evidence that environmental and dietary factors might be involved in this condition.[3] The major focus of diet modifications has been on the dietary fat sources and the effects of specific fats and fatty acids, such as mysteric and palmitic, which have clearly identified effects on elevating blood cholesterol levels.[4-6]

Other dietary factors besides fats and fat sources also influence serum cholesterol levels, and have been studied as to their effects in conjunction with dietary fat modification.[2] One of these factors is the dietary protein source, and soy protein in particular. The earliest indication that dietary protein influenced plasma cholesterol was in studies conducted with rabbits by Meeker and Kesten.[7,8] These researchers showed that the replacement of casein with soy flour in the diet of rabbits significantly reduced the plasma cholesterol level and reduced the degree of atherogenic lesion in the aorta of the rabbits. Carroll and co-workers[9-12] have extended this work and demonstrated that a wide variety of vegetable proteins reduce cholesterol levels in rabbits when compared to a variety of animal protein products. Milk proteins appeared to be the most hypercholesterolemic based on these studies. These responses have been observed in diets which are low in fat, cholesterol-free, and semipurified. The proteins were fed in very purified forms, such as isolated soy proteins, as well as purified sources of animal proteins.

Studies with soy products in human diets have also been conducted over a number of years. Hodges and co-workers[13] published data in 1967 which demonstrated a clear effect of vegetable protein (primarily isolated soy protein) on lowering serum cholesterol levels in men who were hypocholesterolemic. The cholesterol level decreased from 295 to 172 mg/100 ml blood. This effect was observed both with relatively high fat intakes as well as reduced fat intakes of 15% of calories, and did not return to the original levels until the original mixed animal protein diet was fed.

Sirtori and co-workers[14] reported reductions of 21% in plasma cholesterol levels of type II hypercholesterolemic patients in 3 weeks when they were placed on a soy protein diet (Table 1). However, if the comparison is made with the initial period the reduction in cholesterol averages 23%. The reference low-lipid diet with the same calories, fat, and ratio of polyunsaturated to saturated fatty acids produced only an 8.4% reduction in total cholesterol with the same patients. The magnitude of the responses to both diets was somewhat influenced by the order of consumption, but the use of the soybean diet provided a quantum improvement in blood cholesterol over that of the low-lipid diet alone. These reductions in plasma total cholesterol were accompanied by similar reductions in plasma low-density lipoprotein (LDL) cholesterol. The effects were observed both in the presence and absence of additional dietary cholesterol levels. These same Italian researchers also reported on a much larger multicenter study incorporating soy protein into an application-type diet with

TABLE 1
Effect of Dietary Changes on Plasma Total Cholesterol Levels

Order of consumption	Total cholesterol (mg/dl)	Change from initial period (mg/dl)	Percent change from initial
Group I			
Initial period	353		
Low-lipid diet	335	−18	−5.1
Soybean protein diet	257	−96	−27.3
Group II			
Initial period	313		
Soybean protein diet	253	−60	−19.2
Low-lipid diet	276	−37	−11.8

Adapted from Sirtori, C. R., Agradi, E., Conti, F., Mantero, O., and Gatti, E., *Lancet*, 1, 275, 1977.

hypercholesterolemics.[15] With 127 outpatients with type II hyperlipoproteinemia, they were able to show a 23.1% reduction in plasma cholesterol with males and a 25.3% reduction in females with the addition of soy protein to the diet. The responses were greater with nonfamilial hyperlipidemics than familial types.

A series of studies were conducted with isolated soy proteins to evaluate their specific cholesterol-lowering activity in diets. The dietary fats, cholesterol, and energy were similar in diets with and without isolated soy protein. In a study conducted by Goldberg et al.,[16] the effect of replacing animal proteins with isolated soy proteins in identical foods of similar composition was evaluated with mildly hypercholesterolemic patients using the National Institute of Health (NIH) type II diet modification program. Using a crossover design after a 6-week basal period on a type II diet, this study demonstrated a significant reduction in total cholesterol of 3.5% ($p < 0.05$), LDL cholesterol was reduced by 6.0% ($p < 0.015$), and apoprotein B was reduced by 6.3% ($p < 0.05\%$) relative to the diet containing the comparable animal proteins. Although these are relatively small, it should be pointed out that these were effects observed after subjects had been on a modified type II diet. The difference from the base NIH type II diet program, however, was 15.4%. The effect is therefore additive to the dietary fat modification of the typically recommended NIH type II diet program.

In a similar type of study, Wolfe and co-workers[17] also showed a significant reduction in plasma cholesterol levels when an isolated soy protein replaced animal proteins in a NIH type II diet. The degree of hypercholesterolemia was greater in these patients, with initial values averaging 333 mg/dl. Total serum cholesterol was reduced 13% ($p < 0.05$) and LDL cholesterol was reduced by 17% ($p < 0.005$) with the soy protein diet vs. the comparable mixed protein diet.

If the soybean protein diet is compared with the baseline value of 333 mg/dl, then the total cholesterol level was reduced 16%, while total cholesterol was reduced by only 3.6% on the mixed animal protein diet. It is also interesting to observe that there was a 20% reduction in triglycerides when the soy protein diet was fed when compared to the comparable animal protein diet, even though the initial triglyceride levels of the patients were in the normal range.

A study conducted to evaluate the effect of substituting a soy beverage containing 2% butter fat vs. 2% fat cow milk has been reported by Mercer et al.[18] with volunteers who

TABLE 2
Summary of Serum Lipids in Familial Hypercholesterolemic Children Receiving Either a Type II Diet or a Type II Diet with Isolated Soy Protein Added

	No.	Total cholesterol (mg/dl)	LDL cholesterol (mg/dl)	Apoprotein B (mg/dl)
Type II Diet				
Initial	8	312	253	142
2 weeks	8	306	248	155
8 weeks	6	269	209	134
Type II Diet Plus Soy Protein				
Initial	11	336	275	177
2 weeks	11	297	249	144
4 weeks	8	248	203	
8 weeks	5	228	174	122

Summarized from Widhalm, K., in *Nutritional Effect on Cholesterol Metabolism*, Beyen, A. C., Ed., Transmondial, Voorhuzen, 1986.

normally consume milk. This population was totally unscreened for cholesterol levels until they were introduced into the study. Consuming 500 ml of a soy protein beverage made with butter fat vs. consuming 500 ml of 2% fat cow milk did not show an effect of lowering cholesterol levels of the entire study group. However, with individuals who were above the 90th percentile of cholesterol levels and who initially had a cholesterol level averaging 276 mg/dl, a significant reduction in total plasma cholesterol level was observed when the soy drink was consumed. No other diet modification was attempted in this particular study. Therefore, this relatively minor modification in diet protein was able to reduce cholesterol levels in hypercholesterolemic individuals, but showed no apparent effects with the normal population with a more homeostatic level of cholesterol. This observation of lack of effect with normal cholesterolemics has been reported by other workers.[16]

More recently, Widhalm[19] has reported results of adding isolated soy protein to a NIH type II diet of hypercholesterolemic children (Table 2). Rather than substituting for proteins in the diets, he added 20 g of isolated soy protein to these diets for an 8-week period. His results showed a reduction of 32% in total cholesterol and 37% in LDL cholesterol in children with familial hypercholesterolemia on a NIH type II diet with soy protein added. In addition, there was a marked reduction in apo-B-lipoprotein levels by 31%, compared to a 5% reduction in the standard type II diet. These results are similar to the responses observed by Verrillo et al.[20] This latter study (Table 3) evaluated both the effect of substituting soy protein for animal proteins and just adding soy proteins to the diet. Their observations indicated that the effect was observed in both cases, indicating that the soy protein, per se, had some direct benefits in lowering plasma cholesterol. Gaddi and co-workers[21] also have reported on the hypercholesterolemic effects of a soy protein diet with hypercholesterolemic children. Their studies reveal a significant reduction in both total and LDL cholesterol. Total cholesterol was reduced 19.2% after 4 weeks on dietary treatment and 21.8% after 8 weeks on the study.

A recent study by Jenkins et al.[22] evaluating a weight-loss program also suggests that the addition of isolated soy protein to the diet has a cholesterol-lowering effect compared with a comparable casein product. The effect was observed even though the isolated soy protein supplement was added to 2% milk and consumed as two meals per day.

TABLE 3
Plasma Lipid Changes from Substituting and Adding Soybean
Protein to the Diet

	Soybean protein diet			
	Substitution		Addition	
Total cholesterol	mg/dl	Percent change	mg/dl	Percent change
Initial	340		336	
After 16 weeks	240	− 29.5	236	− 29.9
LDL cholesterol, initial	259		243	
After 16 weeks	159	− 38.8	155	− 36.5

Adapted from Verrillo, A., de Teresa, A., Giarusso, P. C., and La Rocca, S., *Atherosclerosis*, 54, 321, 1985.

Results reported by Miyazima et al.[23] found total cholesterol levels were reduced significantly in subjects who had initial values of 226 mg total cholesterol per deciliter. After a 3-week feeding period, the total cholesterol had reduced to 194 ml/dl on a soy protein diet, while remaining at 232 mg/dl on the control diet. This response was obtained with moderately elevated cholesterol levels and are similar to the results by the same researchers[24] in which they showed a reduction of 13% total cholesterol in a group of individuals having initial cholesterol values of only 197 ml/dl.

Danish workers[25] have reported a reduction in LDL cholesterol levels (16%) when a soy protein liquid diet was compared with a similar diet made with casein (milk protein) with normolipidemic subjects (total cholesterol 3.5 to 6.7 mmol/l). This response with normolipidemic individuals may be associated in part with the level of cholesterol (500 mg/ dl) in the diet. A previous study[26] by the same authors using a similar diet which provided dietary cholesterol levels of less than 100 mg/d did not observe a difference between the two protein sources. This suggests that the addition of isolated soy protein to diets with normal cholesterol levels in the 200- to 300-mg/d range would provide a benefit to the consumer.

The body of research developed in the last 20 years clearly indicates that inclusion of soy protein in the human diet can provide additional cholesterol-lowering benefits to individuals who have high blood cholesterol levels. The effects which have been observed in general are greater with individuals who have the higher cholesterol levels, as opposed to individuals who have moderately to slightly elevated cholesterol levels. There have been very few studies that indicate any effects with individuals who have normal cholesterol levels in the range of 140 to 180 mg/dl. There are real benefits to be accrued to individuals having elevated cholesterol levels above 200 mg/dl from including soy protein in the diet, since there are consistent data that suggest this can reduce the risk of heart disease by reducing blood cholesterol levels. The benefit of soy protein are not only in the use of these products to formulate new food products lowering calories, fat, and cholesterol, but they also have demonstrated a benefit of the protein, per se, which is additive to these other dietary modifications. Therefore, while foods can be made lower in fat and cholesterol without soy protein, they would not have the added advantages of soy protein, which can further reduce the risk of heart disease.

REFERENCES

1. Prevention of Coronary Heart Disease, Report of a WHO Expert Committee, Tech. Rep. Ser. No 678, World Health Organization, Geneva, 1982.
2. **Connor, W. E. and Connor, S. J.,** The key role of nutritional factors in the prevention of coronary heart disease, *Prev. Med.,* 1, 49, 1972.
3. **Syme, S. L., Marmot, M. G., Kagan, A., Koto, H., and Rhoads, G.,** Epidemiological studies of coronary heart disease and stroke in Japanese men living in Japan, Hawaii and California, *Am. J. Epidemiol.,* 102, 477, 1975.
4. **Keys, A., Anderson, J. T., and Grande, F.,** Prediction of serum cholesterol responses of man to change in fats in the diet, *Lancet,* 2, 959, 1957.
5. **Hegsted, D. M., McGandy, R. B., Meyers, M. L., and Stare, F. J.,** Quantitative effects of dietary fat on serum cholesterol in man, *Am. J. Clin. Nutr.,* 17, 281, 1965.
6. **Grundy, S. M.,** Monounsaturated fatty acids and cholesterol metabolism: implications for dietary recommendation, *J. Nutr.,* 119, 529, 1989.
7. **Meeker, D. R. and Kesten, H. D.,** Experimental atherosclerosis and high protein diets, *Proc. Soc. Exp. Biol. Med.,* 45, 543, 1940.
8. **Meeker, D. R. and Kesten, H. D.,** Effect of high protein diets on experimental atherosclerosis of rabbits, *Arch. Pathol.,* 31, 147, 1941.
9. **Carroll, K. K. and Hamilton, R. M. G.,** Effects of dietary protein and carbohydrate on plasma cholesterol levels in relation to atherosclerosis, *J. Food Sci.,* 40, 18, 1975.
10. **Carroll, K. K.,** Hypercholesterolemia and atherosclerosis: effects of dietary protein, *Fed. Proc. Fed. Am. Soc. Exp. Biol.,* 41, 2792, 1980.
11. **Huff, M. W. and Carroll, K. K.,** Effects of dietary protein on turnover, oxidation and absorption of cholesterol, and on steroid excretion in rabbits, *J. Lipid Res.,* 21, 546, 1980.
12. **Samman, S., Khosla, P., and Carroll, K. K.,** Effects of dietary casein and soy protein on metabolism of radiolabelled low density apolipoprotein B in rabbits, *Lipids,* 24, 169, 1989.
13. **Hodges, R. E., Krehl, W. A., Stone, D. B., and Lopez, A.,** Dietary carbohydrates and low cholesterol diets: effects on serum lipids in man, *Am. J. Clin. Nutr.,* 20, 198, 1967.
14. **Sirtori, C. R., Agradi, E., Conti, F., Mantero, O., and Gatti, E.,** Soybean protein diet in the treatment of type-II hyperlipoproteinemia, *Lancet,* 1, 275, 1977.
15. **Descovich, G. C., Ceredi, C., Gaddi, A., Benassi, M. S., Mannino, G., Colombo, L., Cattin, L., Fontana, G., Senin, U., Mannarino, E., Caruzzo, C., Bertelli, E., Fragiacomo, C., Noseda, G., Sirtori, M., and Sirtori, C. R.,** Multicenter study of soybean protein diet for outpatients hypercholesterolemic patients, *Lancet,* 2, 709, 1980.
16. **Goldberg, A. P., Lim, A., Kolar, J. B., Grundhauser, J. J., Steinke, F. H., and Schonfeld, G.,** Soybean protein independently lowers plasma cholesterol levels in primary hypercholesterolemia, *Atherosclerosis,* 43, 355, 1982.
17. **Wolfe, B. M., Giovannetti, P. M., Cheng, D. C. A., Roberts, D. C. K., and Carroll, K. K.,** Hypolipidemia effect of substituting soybean protein isolated for all meat and dairy protein in diets of hypercholesterolemic men, *Nutr. Rep. Int.,* 24, 1187, 1981.
18. **Mercer, N. J. H., Carroll, K. K., Giovannetti, P. M., Steinke, F. H., and Wolfe, B. M.,** Effects of human plasma lipids of substituting soybean protein isolate for milk protein in the diet, *Nutr. Rep. Int.,* 35, 279, 1987.
19. **Widhalm, K.,** Effect of diet on serum lipids and lipoprotein in hyperlipoproteinemic children, in *Nutritional Effect on Cholesterol Metabolism,* Beyen, A. C., Ed., Transmondial, Voorhuzen, 1986.
20. **Verillo, A., de Teresa, A., Giarusso, P. C., and La Rocca, S.,** Soybean protein diets in the management of type II hyperlipoproteinemic, *Atherosclerosis,* 54, 321, 1985.
21. **Gaddi, A., Descovich, G. C., Noseda, G., Fragiacomo, C., Nicolini, A., Montanari, G., Vanetti, G., Sirtori, M., Gatti, E., and Sirtori, C. R.,** Hypercholesterolemia treated by soybean protein diet, *Arch. Dis. Child.,* 62, 274, 1987.
22. **Jenkins, D. J. A., Wolever, T. M. S., Spiller, G., Buckley, G., Lam, Y., Jenkins, A. L., and Josse, R. G.,** Hypocholesterolemic effect of vegetable protein in a hypocaloric diet, *Atherosclerosis,* 78, 99, 1989.
23. **Miyazima, E. S., Takeyama, S., Kondo, K., Kagami, A., Suzuki, N., Takeda, N., Ishikawa, T., and Nakamura, H.,** Effect of soy protein substituting diet on hyperlipoproteinemic state, *Nutr. Sci. Soy Prot.,* 3, 90, 1982.
24. **Miyazima, E., Takeyama, S., Tada, N., Ishikawa, T., and Nakamura, H.,** Clinical experiences with soy protein substituted diet on the plasma lipids, *Nutr. Sci. Soy Prot.,* 2, 31, 1981.
25. **Meinertz, H., Nilausen, K., and Faergeman, O.,** Soy protein and casein in cholesterol-enriched diets: effects on plasma lipoproteins in normolipidemic subjects, *Am. J. Clin. Nutr.,* 50, 786, 1989.
26. **Meinertz, H., Faergeman, O., Nilausen, K., Champan, M. J., Goldstein, S., and Lapland, P. M.,** Effects of soy protein and casein in low cholesterol diets on plasma lipoproteins in normolipidemic subjects, *Atherosclerosis,* 72, 63, 1988.

Chapter 21

ANTIATHEROGENIC CHARACTERISTICS OF THE ISOLATED SOYA PROTEIN 500E — CLINICAL EXPERIMENTAL STUDY

Victor A. Tutelyan, Andrei V. Vasilyev, and Li Khva Ren

I. INTRODUCTION

The hypocholesterolemic effect of individual components of food products, and vegetable proteins in particular, is well known.[1,2] Thus, the consumption by rats of proteins obtained from *Cicer arcetinum* brought about a reduction in the level of cholesterol in the serum of the blood and kidney.[3] The inclusion of soya proteins, wheat gluten, and gluten enriched with lysine and threonine in the rations of rats on an atherogenic diet also led to a reduction in the concentration of cholesterol in the blood serum.[4,5] In our studies it was shown that the consumption by patients suffering from an ischemic heart condition of an antiatherosclerotic diet in which 50% of the protein was made up of an isolated soya protein in the course of 2 months led to the normalization of the indicators of lipid metabolism in the blood serum and a drop in the level of activity of lysosomal hydrolysis of thrombocytes (proteinase and phospholipase A) that are potentially capable of participating in the catebolism of lipoproteins.[6]

Inasmuch as hypercholesterolemia is inevitably linked with the development of the atherosclerotic process,[7] there is every reason to presume that the hypocholesterolemic effect of individual vegetable proteins will be accompanied by an antiatherogenic (prophylactic) effect. In this respect, taking into account the possibility of large-scale use and economic feasibility, the most attractive are the protein concentrates and isolates of vegetable origin, in particular, isolated soya proteins.[8]

For evaluating the potential antiatherogenic characteristics of the isolated soya protein 500E (produced by Protein Technologies International, U.S.), a group of indicators characterizing the condition of lipid metabolism and reflecting the development of the process of atherogenesis were used: (1) the total cholesterol content in the blood serum; (2) the cholesterol content in the composition of the circulating immune complexes (CIC), since there is information about the atherogenic nature of low-density lipoproteins, linked, for example, with the immunoglobulin group G;[9,10] (3) the activity of the lysosomal proteinase and phospholipase A of thrombocytes capable of taking part in the modification of low-density lipoproteins; and (4) also the first layer of subendothelial atherosclerotic cells of the intima of the human aorta. In the latter culture, the basic atherosclerotic properties of the cells are preserved, and this makes it possible to consider a culture of intima cells as a convenient adequate model for studying the basic manifestations of atherosclerosis and for studying the antiatherogenic effect of various preparations.[11,12]

II. METHODOLOGY

Immediately before, and at 2 and 4 h after taking, on an empty stomach, 50 g dry milk protein containing caseins and lactalbumins in a ratio of 60:40 or 50 g isolated soy protein 500E, 20 ml of blood were taken from the elbow vein of patients suffering from postcoronary atherosclerosis. The patients were men aged 50 to 62 years with a body weight of 70 to 80 kg being treated in the cardiovascular division of the Therapeutic Nutrition Clinic of the Institute of Nutrition of the U.S.S.R Academy of Medical Sciences. By means of gradual centrifugation, thrombocytes were extracted from 10 ml of the blood, using as the separating medium a 6% solution of ficol with 76% verografin.[13] The thrombocytes that are obtained are homogenized in 0.25 M sucrose containing 1 mm EDTA. In the homogenates of the isolated thrombocytes the activity of the lysosomal enzymes were determined: the cathepsins B (EC 3.4.22.1) and C (EC 3.4.14.1) and the phospholipases A_1 (EC 3.1.1.32) and A_2 (EC 3.1.1.4).

The activity of the cathepsins B and C was determined by fluorescence spectroscopy using *N*-benzoyl-DL-arginine-β-naphthylamide and glycyl-L-phenylalanine-β-naphthylamide, respectively (Sigma, U.S.), as the substrate. The activity of phospholipases A_1 and A_2 was determined by radiometry, using 1-acyl-2 (1-^{14}C)-linoleil-3 Sn-glycero phosphorylcholine synthesized by us as the substrate.[14-16]

From the remaining blood the serum total cholesterol content and the cholesterol in the circulating immune complexes[17] were measured. The remaining quantity was used to study the cells of the intima of the human aorta using the culture techniques.

A water-alcohol extract of the isolated soya protein 500E was obtained by means of extracting a sample of the soya protein with 70% ethyl alcohol (100:1, volume/mass) at 37°C during 20 h with constant agitation.

The cells of the subendothelial intima of the human aorta were obtained aseptically from fresh autopsy material taken 1 to 3 h after sudden death from an unhospitalized man between 40 and 60 years old. The cells were isolated by the dispersion of the atherosclerotic plaque with 0.15% collagenase, type IV (Sigma, U.S.), dissolved in a Ca^{2+}, Mg^{2+} phosphate buffer (PBS), pH 7.3, containing 0.1% glucose, 25 mM HEPES, 10% embryonal calf serum, 2.5 mg/ml kanamycin, and 25 mg/ml fungisone (GIBCO® Europe, Paisley, U.K.) at a temperature of 37°C in an atmosphere of CO_2/air (5:95) saturated with water. The medium was changed every 3 d. On the seventh day of incubation, 10 ml extract of the isolated soya protein was added to the medium. After 24 h the cells were washed and their total cholesterol content and the degree of inclusion of ^{3}H-thymidine in DNA were determined.

The results were processed by the method of dispersion analysis using the statistical software BMDP IV.

III. RESULTS AND DISCUSSION

A comparative evaluation of the effect of proteins of animal and vegetable origin on the indicators of lipid metabolism, the activity of lysosomal hydrolysis in the thrombocytes, and the nature of the action of the serum on affected cells of the intima provide a basis for identifying isolated soya protein as a product manifesting antiatherogenic characteristics. Within 2 h after the isolated soya protein 500E was taken by patients, a reduction of 50% is noted in the activity of phospholipase A_1 in the thrombocytes, the cholesterol content in the circulating immune complexes is lowered, and the concentration of intracellular cholesterol in the cultivated cells of the intima is reduced by 29%. Within 4 h of study, a decline in the activity of the lysosomal proteinases of the thrombocytes — cathepsins B and C by 20 and 24%, respectively — is observed (Table 1). It is significant that similar changes were not revealed when milk proteins were used as the food source. Since the total cholesterol content in the blood serum did not change, but, nevertheless, a reduction of cholesterol in the composition of the circulating immune complexes was found in combination with the antiatherosclerotic effect of the serum in the culture of the cells, apparently one can hypothesize that the antiatherogenic action of isolated soya protein 500E is expressed through factors in the serum other than cholesterol itself. Together with this, the results concerning the direct antiatherosclerotic effect of the water-alcohol extract of the protein extract 500E on the cells of the intima (Table 2), which reduced the intracell cholesterol content by 30% and the suppressed cell proliferation by 24%, provides a basis for linking the biological effect of this protein product with its amino acid composition. The available information suggests possible effect of free tyrosine, valine, and the ratios of leucine-isoleucine on the metabolism of cholesterol and its content in the blood serum.[18,19] Thus, there is every reason to consider the isolated soya protein 500E as a necessary component of antiatherosclerotic diets, which should be taken into account in the dietetic treatment of cardiovascular diseases.

TABLE 1
Hypolipidemic and Antiatherogenic Properties of the Isolate Soya Protein 500E

Characteristics studied	Milk proteins			Isolated soya protein 500E		
	Before taking	After 2 h	After 4 h	Before taking	After 2 hr	After 4 h
Activity of lysosomal hydrolysis in the thrombocytes (μmol/min for 1 g protein)						
Cathepsin B	0.239 ± 0.009	0.224 ± 0.018	0.201 ± 0.015	0.272 ± 0.019	0.244 ± 0.020	0.218 ± 0.016[a]
Cathepsin C	2.23 ± 0.24	2.37 ± 0.24	1.80 ± 0.15	3.16 ± 0.29	2.90 ± 0.23	2.40 ± 0.12[a]
Phospholipase A$_1$	1.21 ± 0.11	1.49 ± 0.12[a]	1.07 ± 0.10	1.92 ± 0.18	0.75 ± 0.11[a]	1.87 ± 0.12[a]
Phospholipase A$_2$	1.60 ± 0.18	1.98 ± 0.21	1.53 ± 0.12	1.75 ± 0.14	1.86 ± 0.20	1.72 ± 0.16
Total serum cholesterol (mg/100)						
	228.0 ± 30.0	223.0 ± 21.0	207.0 ± 18.0	290.0 ± 31.0	270.0 ± 25.0	286.0 ± 29.0
Quantity of cholesterol in the immune complexes containing lipoproteins (μg/ml)						
	14.3 ± 1.5	20.8 ± 2.2[a]	11.6 ± 1.7	18.1 ± 1.2	12.8 ± 0.9[a]	12.4 ± 1.1[a]
Cholesterol content in atherosclerotic cells of the intima of the human aorta when introduced in a culture of the serum of diseased persons (μg/ mg protein)						
	365.0 ± 35.8	295.3 ± 35.0	335.8 ± 33.2	410.0 ± 30.0	290.3 ± 23.8[a]	286.2 ± 22.6[a]

Note: The average data (mean ± SE) from five experiments are presented.

[a] $p < 0.05$

TABLE 2
The effect of a Water-Alcohol Extract of the Isolated Soy Protein 500E on the Accumulation of Cholesterol and the Proliferation of Atherosclerotic Cells of the Human Intima

Cholesterol content (μg/mg of protein)		Uptake of ^{3}H-thymidine (Pulse/mg of protein)	
Control	Experiment	Control	Experiment
300 ± 29	231 ± 23 ($p < 0.05$)	115 ± 10	88 ± 9 ($p < 0.05$)

REFERENCES

1. **Jaya, T. V., Mengheri, E., and Scarino, M. L. et al.,** Evaluation of hypocholesterolemic effect of faba bean protein concentrate on rats fed a high-fat-cholesterol diet, *Nutr. Rep. Int.,* 23, 55, 1981.
2. **Sugano, M.,** Hypocholesterolemic effect of plant protein in relation to animal protein mechanism of action, *Curr. Top. Nutr. Dis.,* 8, 54, 1983.
3. **Kowsalya, S. M. and Kantharaj, U. M.,** Effect of bengal gram (cicerarietinum) proteins and lipids on serum and liver cholesterol level in rats, *J. Food Sci. Technol.,* 22, 54, 1985.
4. **Leelamma,., Menon, P. K. G., and Kurup, P. A.,** Nature and quantity of dietary protein and metabolism of lipids in rats fed normal and atherogenic diet, *Indian J. Exp. Biol.,* 16, 29, 1978.
5. **Mokady, S. and Lienen, J. R.,** Effect of plant proteins and cholesterol metabolism in growing rats fed atherogenic diets, *Ann. Nutr. Metab.,* 26, 138, 1982.
6. **Vasilyev, A. V., Meshcheryakova, V. A., and Pogozheva, A. V. et al.,** Effect of dietotheraphy on the activity of blood platelet lysosomal hydrolysis in patients with hyperlipidemia and ischemic heart disease, *Patol. Fiziol. Eksp. Ter.,* 4, 71, 1987.
7. **Klimov, A. N. and Nikulcheva, N. G.,** *Lipoproteidy, Dislipoproteidemii I Ateroskleroz* (Lipoproteins, Dyslipoproteinemia and Atherosclerosis), Meditsina, Leningrad, 1984.
8. **Lo, G. S., Evans, R. H., and Phillips, K. S. et al.,** Effect of soy fiber and soy protein on cholesterol metabolism and atherosclerosis in rabbits, *Atherosclerosis,* 64, 47, 1987.
9. *Immunoreaktivnost' I Ateroskleroz* (Immunoreactivity and Atherosclerosis), Klimov, A. N., Ed., Meditsina, Leningrad, 1986.
10. **Klimov, A. N., Denisenko, A. D., and Popov, A. V. et al.,** Lipoprotein-antibody immune complexes; their catabolism and role in foam cell formation, *Atherosclerosis,* 58, 1, 1985.
11. **Orekhov, A. N. and Kosykh, V. A.,** Functional and metabolic characterization normal and atherosclerotic aorta cells of man, in *Stenka Sosudov v atero- i trombogeneze* (vessel walls in atherogenesis and thrombogenesis), Chazova, E. I. and Smirnova, V. N., Eds., Meditsina, Moscow, 1983, 53.
12. **Orekhov, A. N., Tertov, V. V., and Kudryashov, S. A. et al.,** Primary culture of human aortic intima cells as a model for testing antiatherosclerotic drugs, *Atherosclerosis,* 60, 101, 1986.
13. **Sokovnina, Ya. M., Pestina, T. I., and Tentsova, I. A. et al.,** Adenosine deaminase and purine nucleosidephosphorylase of blood platelets in different blood diseases, *Vestn. Akad. Med. Nauk SSSR,* 8, 75, 1984.
14. **Vasilyev, A. V., Kapelevich, T. A., and Tutelyan, V. A.,** Spectrofluorimetric procedures for estimation of lysosomal cathepsins A, B, C and C activities, *Vopr. Med. Khim.,* 3, 127, 1983.
15. **Robertson, A. J. and Lands, W. R. M.,** Positional specificities in phospholipid hydrolysis, *Biochemistry,* 804, 1962.
16. **Stoffel, W. and Trabert, U.,** Studies on the occurrence and properties of lysosomal phospholipases A_1 and A_2 and the degradation of phosphatidic acid in rat liver lysosomes, *Hoppe-Seyler's Z. Physiol. Chem.,* 350, 836, 1969.
17. **Szondy, H., Horvath, M., and Mezey, Z. et al.,** Free and complexed antilipoproteins antibodies in vascular disease, *Atherosclerosis,* 49, 69, 1983.
18. **Carroll, K. K.,** Effects of dietary proteins and amino acids on serum cholesterol levels, in *New Trends Pathophysiol. and Ther. Large Bowel,* Proc. Int. Symp., Bologna, April 7 to 8, 1983, Miglioli, M., Phillips, S. F., and Barbara, L., Eds., Elsevier Science Publishers B. V., Amsterdam, 1983, 293.
19. **Jacques, H., Deshaies, Y., and Savoie, L.,** Relationship between dietary proteins, their in vitro digestion products and serum cholesterol in rats, *Atherosclerosis,* 61, 89, 1986.

Chapter 22

DIET THERAPY OF OBESE PATIENTS WITH HYPERTENSION, ISCHEMIC HEART DISEASE, OR HYPERLIPOPROTEINEMIA WITH ISOLATED SOY PROTEIN FOODS

Michael N. Volgarev, Victor A. Tutelyan, Michael A. Samsonov, Vadim G. Vysotsky, Galina R. Pokrovskaya, Alexander S. Vitollo, Irina S. Zilova, Natalya P. Shimanovskaya, and Olga M. Kalamkarova

Previous joint studies between the Institute of Nutrition, U.S.S.R. Academy of Medical Sciences, and Protein Technologies International demonstrated a good tolerance by patients with second- or third-degree obesity and hypercholesterolemia for a number of foodstuffs with isolated soy protein (five types of products), contributing the basis for a low-calorie (1300 kcal per day) diet, the reducing effect of this diet on body weight, and its pronounced hypocholesterolemic effect.[1] This previous study served as a basis to develop formulations for a wider variety of foodstuffs made with isolated soy protein belonging to different food groups and subsequent testing of their clinical efficiency.

To conduct these trials, Protein Technologies International, VNIKIMD (Milk Institute), and VNIKMI (Dairy Institute) produced 12 types of combination food products with isolated soy protein (wheat and oat porridge, chicken and fish "nuggets", two types of pasta, four types of sausages and frankfurters, two types of beverages) which were used to make a test diet (7-d menu) with reduced energy value (about 1700 kcal/d). The same types of products, but without containing isolated soy protein (cow milk was used instead of soy-based beverages), served as a basis to make a control diet (7-d menu) with the similar test energy value.

The 35-d study consisted of a background (1 week) and test (4 weeks) periods. In the background period, all patients (30 males) received the regular therapeutic antiatherosclerotic diets used in the Institute of Nutrition clinic with the three types of heart diseases. In the test period, these patients were divided into 2 groups of 15 patients each. The first group received test diet and the second group received control diet. The estimated chemical composition and the daily menu of all diets used are given in Tables 1 and 2, respectively. All patients received one Undevit multivitamin tablet daily. Each group (test and control) consisted of three subgroups of five patients, each with the following types of disease:

- first subgroup — patients with ischemic heart disease (IHD)
- second subgroup — patients with hyperlipidemia (HLP)
- third subgroup — patients with hypertension disease (HD)

The age characteristics of the patients examined, their initial body weight, and body mass index are summarized in Table 3 (mean ± SE). From the data described in Table 3, it follows that there were no differences observed according to age distribution between test and control groups of patients within each specific disease group. All patients had excessive body weight, since only those patients whose Ketle index was above 25 were chosen.

The evaluation of the obesity type was based on the conventional assumption according to which the value of the index within 25.0 to 29.9 units corresponds to the initial obesity, clinical obesity to 30.0 to 39.9 units, and 40.0 units and above to the severe forms of obesity. It follows from this classification that obesity in the patients in the test and control groups conformed with the clinical type with a somewhat higher value for the Ketle index in patients with hypertension.

TABLE 1
Average Daily Chemical Composition and Energy Value of Test Diets

	Background diet	Test diet (with soy isolate)	Control diet (without soy isolate)
Protein (g)	108	85	76
Animal	60	19.3	49
Vegetable, total	48	65.7	27
Isolated soy protein		44.5	
Fats (g)	81	62	71
Animal	56	33	51
Vegetable	25	29	20
Cholesterol (mg)	356	86	296
Carbohydrates (g)	321	206	188
Refined	35	40	35
Calories (kcal)	2545	1722	1695

During the whole period of study, the patients were under the controlled conditions of the inpatient department of the hospital for 24 h with a regular regimen of daily activity and rest. The medical examination was performed daily to measure the following conditions: (1) how complete was the diet consumed, (2) its tolerance and feeling of repletion as well as body temperature, (3) arterial pressure, and (4) pulse frequency. The blood analysis was performed during the first and fifth weeks and the content of urea and creatinine in serum and their excretion in the urine were evaluated during the first, third, and fifth weeks of the study.

Lipid metabolism status was evaluated by the determination of the total cholesterol concentration in blood serum (TCH; norm 3.0 to 7.10 mmol/l), beta-cholesterol (BCH; norm 0.93 to 1.96 mmol/l), and triglycerides (TG; norm 0.59 to 1.77 mmol/l) using the biochemical analyzer FP-90 (Finland). The content of prebeta-cholesterol (PBCH; norm 0.26 to 0.79 mmol/l) was evaluated by the calculation method, multiplying the amount of TG by 0.44 ratio, and alpha-cholesterol (ACH; norm 1.78 to 6.00 mmol/l) by subtracting the sum of BCH and PBCH from the TCH value. The relationship of atherogenic classes of lipoproteins cholesterol (beta- and prebeta-lipoproteins to nonatherogenic alpha-lipoproteins cholesterol = coefficient of "atherogenicity" [CA]) was calculated according to the formula (norm − 3.0):

$$CA = \frac{TCH\text{-}ACH}{ACH}$$

The status of hemostats and the blood coagulation system was evaluated by the determination of the Queen prothrombin index[2] (norm 82 to 87%), the fibrinogen content by the Rubberf method (norm 200 to 400 mg/%, and fibrinolytic activity by the Kovalskyi method[3] (norm 120 to 180 min). The content of glucose in blood serum was studied (norm 3.4 to 5.6 mmol/l), as well as total protein (norm 60 to 80 g/l) using the biochemical analyzer FP-90, uric acid (norm for males 0.12 to 0.46 mmol/l) by the Morimint-London method,[4] albumin (norm 37 to 52 g/l) using a Beringer-Mannheim (West Germany) kit, creatinine and urea using a Lahema (Czechoslovakia) kit; the latter two values were also evaluated in the daily urine (norm for creatinine in urine 4.4 to 17.6 mmol/d, in serum 44 to 176 mmol/l; norm for urea: in urine 333 to 583 mmol/d, in serum 2.50 to 8.32 mmol/l).

Urinary excretion of total nitrogen as well as urea and creatinine were studied for the last 5 d of the first, third, and fifth weeks. The status of protein nitrogen balance and digestibility were evaluated only in IHD patients in test and groups for the last 5 d of the

TABLE 2A
Typical 1-d Menu of the Background Diet (g/d)

Menu	Amount	Protein	Fats	Carbohydrates
Breakfast 1				
Boiled meat	55	18.49	2.58	0.84
Cucumber salad	170	1.36	9.99	4.42
with vegetable oil	10			
Coffee	130	1.4	1.6	2.35
and milk	50			
Rice porridge butter	250	3.84	8.51	40.21
	10			
Morning Snack				
An apple	100	0.4	—	9.8
Lunch				
Soup of assorted vegetables	500	4.09	6.48	24.02
with sour cream, vegetable oil	5			
Boiled meat	55	18.49	2.58	0.84
Stewed beet	150			
with vegetable oil	10	2.46	9.99	14.92
Apple beverage	180	0.3	—	7.35
Afternoon Snack				
An apple	100	0.4	—	9.8
Dinner				
Boiled fish	100	23.03	9.11	1.06
Boiled potatoes	200			
with vegetable oil	10	4.48	10.80	36.42
Vermicelli	200			
with cheese	30	10.77	14.36	34.89
Tea	180	1.58	1.6	2.38
and milk	50			
Evening Snack				
Yogurt (kefir)	180	5.04	1.08	7.38
For the Whole Day				
Rye bread	100	5.51	1.0	41.0
Wheat bread	100	7.6	0.9	49.9
Sugar	35	—	—	34.93
Total	**2457 kcal**	**109**	**81**	**323**

first and fifth weeks of their hospitalization. The determination of total nitrogen in the diets, daily urine and feces combined samples for 5 d was done by the Kjeldahl method using the Kjeltech® auto-1030 from Techator (Sweden).

The isolated soybean protein tolerance data analysis (Table 4) showed that in general the test diet completely satisfied the patients by its taste qualities. Preference was given to the experimental diet vs. the conventional hypocaloric and antiatherosclerotic (based on

TABLE 2B
Typical 1-d Menu of the Test Diet with Isolated Soy Protein (g/d)

Menu	Amount	Protein	Fats	Carbohydrates
Breakfast 1				
Pasta with sauce	60 (dry)	14.34	0.9	38.1
Cucumber salad	170			
with vegetable oil	5	1.36	4.99	4.42
Chocolate beverage	250	8.83	5.98	31.28
Morning Snack				
A fresh apple	100	0.4	—	9.8
Lunch				
Mushroom soup	27 (dry)	12.0	1.0	7.99
"Zdorovje" sausage	100	13.8	15.0	0.8
Stewed beets	150			
with vegetable oil	5	2.46	9.99	14.92
Apple beverage without sugar	180	0.3	—	7.35
Afternoon Snack				
A fresh apple	100	0.4	—	9.8
Dinner				
Fish nugget	80	13.28	8.72	11.68
Stewed cabbage with vegetable oil	200	1.42	10.05	13.02
Tea	180	0.18	—	0.03
Evening Snack				
Vanilla beverage	250	8.33	5.68	20.0
For the Whole Day				
Wheat bread	100	7.6	0.9	49.7
A lemon	60	—	—	0.18
Total	**1738kcal**	**85**	**58**	**219**

traditional products) by 86% of the patients studied. Of this group, 60% of the patients associated their positive attitute towards the test diet with an excellent taste quality and 26% due to their conviction in the diet usefulness. All the patients without exception indicated a good repletion feeling during the test diet period vs. the conventional hypocaloric diet that they had used previously. An additional positive clinical effect of products with isolated soy protein was a normalizing effect on intestinal function that eliminated constipation in one third of cases. It must be noted also that while using calorie-reduced diets, an inclination to stool delay is a common problem. In a number of cases, however, a negative attitude towards some meals with isolated soy protein was observed: 20% of patients noted an unpleasant soy taste in porridge, 66% of patients in the vanilla beverage.

Body weight dynamics in the study are shown in Table 5. The loss of body weight was practically identical in both test and control groups. For the three subgroups (IHD, HLP, HD) the weight loss was respectively as follows: 5.8%, 6.3%, and 9.4% for the patients in the test group and 7.9%, 6.6%, and 12% for those in the control group. The Ketle index

TABLE 2C
Typical 1-d Menu of the Control Diet (g/d)

Menu	Amount	Protein	Fats	Carbohydrates
Breakfast 1				
Vermicelli with cheese	180/60	18.65	22.55	34.89
Cucumber salad	170	1.64	2.0	4.74
with sour cream	10			
Milk	180	5.04	5.76	8.46
Morning Snack				
A fresh apple	100	0.4	—	9.8
Lunch				
Vegetarian cabbage soup	250	2.175	3.16	14.81
($^1/_2$ portion)				
Sausage	100	13.8	15.0	0.8
Stewed beet	150	2.46	4.99	14.92
with vegetable oil	5			
Apple beverage	180	0.3	—	7.35
Afternoon Snack				
A fresh apple	100	0.4	—	9.8
Dinner				
Boiled fish	100	23.03	4.12	1.06
Stewed cabbage with vegetable oil	180	0.18	—	0.07
Evening Snack				
Milk	180	5.04	5.76	8.46
For the Whole Day				
Wheat bread	100	7.6	0.9	49.7
Lemon	60	—	—	0.18
Total	**1706 kcal**	**82**	**74**	**178**

TABLE 3
Characteristics of the Subjects[a]

Disease	Test group			Control group		
	Age (years)	Body weight (kg)	Ketle index (units)	Age (years)	Body weight (kg)	Ketle index (units)
IHD	50.0 ± 4.8	94.2 ± 8.6	31.0 ± 3.0	54.0 ± 5.1	87.4 ± 7.2	31.0 ± 3.0
HLP	57.0 ± 5.0	86.8 ± 7.9	32.0 ± 3.0	55.4 ± 5.2	90.0 ± 8.7	31.0 ± 2.0
HD	40.8 ± 4.5	116.0 ± 13.1	37.0 ± 4.0	46.6 ± 4.3	107.0 ± 8.9	34.0 ± 3.0

Note: IHD, ischemic heart disease; HLP, hyperlipidemia; HD, hypertension disease.

[a] Mean ± SE.

TABLE 4
Tolerance Evaluation of the Diet with Soy Protein
Isolate

Evaluation criteria	Number of cases (%)
Evaluation of taste qualities:	
Patient is fully satisfied	33
Unpleasant aftertaste in porridge	20
Unpleasant aftertaste in vanilla beverage	66
Preferred diet with soy protein	
Isolate vs. conventional hypocaloric antiatherosclerotic (Ap):	
Due to good taste	60
Due to conviction in the diet usefulness	26
Good repletion feeling	100
Changes in stool:	
Elimination of constipation	33
Liquid stool	20
Without changes	47

TABLE 5
Body Weight (Mean ± SE) in Patients of Test and Control Groups
Before and After Treatment

	Test group		Control group	
Patients subgroups	Body weight (kg)	Ketle index (units)	Body weight (kg)	Ketle index (units)
IHD before treatment	94.2 ± 8.6	31.0 ± 3.0	87.4 ± 7.2	31.0 ± 3.0
After treatment	88.7 ± 7.8	29.0 ± 2.0	80.5 ± 6.8	28.0 ± 2.0
Difference in percent	−5.8	−7.0	−7.9	−10.0
HLP before treatment	86.8 ± 7.9	32.0 ± 3.0	90.0 ± 8.7	31.0 ± 2.0
After treatment	81.3 ± 6.5	26.0 ± 2.0	84.0 ± 7.5	29.0 ± 3.0
Difference in percent	−6.3	−15.0	−6.6	−7.0
HD before treatment	116.0 ± 13.1	37.0 ± 4.0	107.0 ± 8.9	34.0 ± 3.0
After treatment	106.0 ± 9.6	33.0 ± 3.0	94.0 ± 9.1	30.0 ± 3.0
Difference in percent	−9.4	−11.0	−12.0	−12.0

Note: IHD, ischemic heart disease; HLP, hyperlipidemia; HD, hypertension disease.

in patients with IHD and HLP was reduced in both groups to the level characteristic for the initial form of obesity (below 29.9 units). In patients with HD who had a more pronounced rate of obesity, in spite of the considerable weight loss during the treatment period, the Ketle index did not reach a similar level, remaining within the value limits attributed to the clinically expressed forms of obesity.

One of the objectives of this study was the evaluation of the efficiency of soy correlated effect on the disturbed lipid metabolism. As the analysis of the results shows (Tables 6 and 7), interrelations of atherogenic and antiatherogenic cholesterol fractions in blood serum varied within rather a wide range in both the test and control groups.

It is interesting to note that with pronounced hypercholesterolemia (in patients with HLP), the normalizing effect of a diet which included isolated soy protein was obvious, in particular CA-integral index determining atherogenesis activity by the end of treatment was reduced by 31% in the test group, while in the control group it reduced only by 13%. A positive shift in the indices of lipid metabolism with the use of soy was provided by the

TABLE 6

Indices of Lipid Metabolism in Blood Serum in Ischemic Heart Disease Patients, Receiving Diets with Isolated Soy Protein (Test Group) and Without It (Control Group)

Indices (mmol/l)	Test group			Control group		
	Before treatment	After treatment	Difference (%)	Before treatment	After treatment	Difference (%)
TCH	6.8 ± 0.47	4.84 ± 0.49[a]	−29	6.73 ± 0.42	4.54 ± 0.41	−32
ACH	1.16 ± 0.09	0.88 ± 0.08	−14	0.92 ± 0.08	0.92 ± 0.08	0
TG	3.42 ± 0.14	0.80 ± 0.06[a]	−76	2.78 ± 0.13	0.68 ± 0.05[a]	−75.5
CH-LDL	4.14 ± 0.32	3.60 ± 0.31	−13	4.59 ± 0.38	3.33 ± 0.31	−25
CH-VLDL	1.50 ± 0.05	0.35 ± 0.04[a]	−76	1.22 ± 0.07	0.30 ± 0.03[a]	−75
CA (units)	4.86 ± 0.38	4.50 ± 0.45	−7.5	6.30 ± 0.40	3.90 ± 0.41	−38

Note: TCH, total cholesterol; ACH, alpha-cholesterol; TG, triglycerides; CH-LDL, cholesterol in low-density lipoproteins; CH-VLDL, cholesterol in very low-density lipoproteins; CA, atherogenicity coefficient.

[a] $p < 0.05$.

TABLE 7

Indices of Lipid Metabolism in Blood Serum in Patients with HLP and HD of Test and Control Groups

Indices (mmol/l)	Test group			Control group		
	Before treatment	After treatment	Difference (%)	Before treatment	After treatment	Difference (%)
Hyperlipoproteinemic Patients						
TCH	7.83 ± 0.61	5.39 ± 0.51	−31	7.65 ± 0.57	4.97 ± 0.46[a]	−35
ACH	1.29 ± 0.09	1.20 ± 0.09	−7	1.58 ± 0.11	1.13 ± 0.09	−23
TG	2.11 ± 0.18	1.34 ± 0.12[a]	−36.5	2.04 ± 0.18	1.19 ± 0.10[a]	−41.6
CH-LDL	5.61 ± 0.51	3.72 ± 0.36[a]	−34	4.17 ± 0.40	3.32 ± 0.30	−20
CH-VLDL	0.93 ± 0.06	0.59 ± 0.05[a]	−36	0.89 ± 0.09	0.52 ± 0.04[a]	−41
CA (units)	5.07 ± 0.50	3.50 ± 0.31[a]	−31	3.87 ± 0.40	3.39 ± 0.34	−13
Hypertensive Patients						
TCH	5.03 ± 0.43	4.31 ± 0.39	−23	4.73 ± 0.32	4.45 ± 0.35	−6
ACH	0.82 ± 0.07	0.84 ± 0.08	+2	0.86 ± 0.07	0.90 ± 0.08	+5
TG	2.63 ± 0.10	1.57 ± 0.12	−40	1.56 ± 0.11	1.47 ± 0.14	−6
CH-LDL	3.05 ± 0.29	2.78 ± 0.26	−9	3.41 ± 0.29	2.79 ± 0.25	−18
CH-VLDL	1.16 ± 0.05	0.69 ± 0.06	−41	0.56 ± 0.04	0.48 ± 0.06	−14
CA (units)	5.13 ± 0.48	4.10 ± 0.39	−20	4.50 ± 0.40	3.90 ± 0.30	−13

Note: TCH, total cholesterol; ACH, alpha-cholesterol; TG, triglycerides; CH-LPLD, cholesterol in low-density lipoproteins; CH-LPVLD, cholesterol in very low-density lipoproteins; CA, atherogenicity coefficient.

[a] $p < 0.05$.

significant reduction in total cholesterol, CH-LDL, and CH-VLDL vs. the control group, maintaining the stable initial level of CH-HDL (Table 7).*

In those cases where the initial cholesterol level did not exceed the limits of normal indices (in patients with IHD, HD), convincing evidence of an advantage of the diet with soy in terms of normalizing lipid indices was not obtained. As it is seen from Table 6, in

* CH-LDL, low density lipoprotein cholesterol; CH-VLDL, very low density lipoprotein cholesterol; CH-HDL, high density lipoprotein cholesterol.

TABLE 8
Indices of Blood Coagulating and Anticoagulation Systems Before and After Dietary Treatment

	Indices		
Groups and sampling period	Fibrinogen (mg %)	Fibrinolytic activity (min)	Prothrombin (%)
Ischemic heart disease patients			
Test I	346.0 ± 29.5	192.0 ± 15.7	85.0 ± 7.5
II	324.0 ± 30.8	202.0 ± 17.8	88.0 ± 7.8
Control I	399.4 ± 32.3	188.0 ± 16.7	91.0 ± 9.2
II	328.6 ± 31.4	202.0 ± 19.8	84.0 ± 8.0
Hyperlipidemia patients			
Test I	399.6 ± 32.3	198.0 ± 18.1	91.0 ± 8.9
II	399.6 ± 34.5	202.0 ± 18.86	83.0 ± 7.9
Control I	315.0 ± 29.8	204.0 ± 18.68	6.0 ± 8.3
II	355.2 ± 30.4	198.0 ± 19.68	5.0 ± 8.1
Hypertension disease patients			
Test I	382.0 ± 34.7	208.0 ± 18.67	5.4 ± 7.2
II	310.0 ± 30.5	200.0 ± 18.1	80.0 ± 7.1
Control I	324.0 ± 28.91	84.0 ± 15.9	85.0 ± 7.6
II	355.0 ± 32.4	150.0 ± 13.0	82.0 ± 7.8

Note: I = before treatment, II = after treatment.

patients with IHD in the control group the CA index was reduced during treatment by 38%, while in the test group it was reduced only by 7.5%. Meanwhile, atherogenic fractions of lipoproteins in both the test and control groups reduced equally. In the control group, however, no reduction in ACH concentration was noted, while in patients receiving isolated soy protein it was reduced by 13%.

In patients with HD (Table 7) where hypercholesterolemia was absent, CA reduction did not vary in the test and control groups. However, while applying diet with soy, the TCH level was reduced more abruptly by 23%, but only 6% in the control group, and its content within CH-VLDL was reduced by 41%, while in control it was reduced by 14%.

The TG level was significantly reduced equally in the test and control groups in all three types of heart disease, except for the control group of patients with HD, where the TG content was reduced only by 6%.

In general, it can be concluded that there were no clear advantages as related to the hypolipidemic effect of isolated soy protein in this study. The fact that the reduction in the atherogenic properties of lipids took place mainly at the expense of cholesterol in very low-density lipoproteins and TG demonstrates a decisive factor in calorie reduction of diets vs. protein quality content. At the same time, the normalizing effect of soy proteins on this index was confirmed with patients having hypercholesterolemia.

Parameters evaluating the status of blood coagulating and anticoagulating systems during diet therapy (Table 8) showed a tendency towards reduction in blood-coagulating properties, but significant differences in test and control groups were not observed.

Table 9 shows the dynamics of changes in blood glucose. During diet therapy, a reduction in glucose level was observed in practically all patients studied. The mean value of this index did not change in a group of patients with HLP. In general, there were no true differences in the degree of glycemia reduction after fasting in both test and control groups.

The data in Table 10 show the absence of an effect of the test and control diets for a 4-week period of diet therapy on the content of the total protein and albumin in blood serum of the patients. It should be noted that for both groups there was a tendency towards reduction of creatinine concentration in the test period which is probably associated with the lower

TABLE 9
Blood Glucose Levels Before and After Treatment

Groups and sampling period	Blood glucose (mmol/l)
Ischemic heart disease patients	
Test I	7.12 ± 0.68
II	6.50 ± 0.55
Control I	5.20 ± 0.60
II	5.00 ± 0.40
Hyperlipidemia patients	
Test I	5.83 ± 0.49
II	5.33 ± 0.43
Control I	5.30 ± 0.50
II	5.40 ± 0.60
Hypertension disease patients	
Test I	5.85 ± 0.46
II	4.43 ± 0.39
Control I	5.02 ± 0.40
II	4.10 ± 0.30

Note: I = before test, II = after test period.

TABLE 10
Indices of Protein Metabolism

Groups and sampling period	Total protein (g/l)	Albumin (g/l)	Creatinine (μmol/l)	Urea (mmol/l)	Uric acid (mol/l)
Ischemic heart disease patients					
Test I	76.0 ± 20.9	46.0 ± 1.2	128.0 ± 11.8	4.82 ± 0.95	0.25 ± 0.02
II	74.5 ± 7.1	42.2 ± 1.8	116.4 ± 1.9	8.71 ± 0.41	0.32 ± 0.03
Control I	74.1 ± 6.8	44.1 ± 2.1	127.5 ± 4.1	4.51 ± 0.51	0.28 ± 0.02
II	74.9 ± 6.4	40.2 ± 2.8	111.7 ± 14.1	8.27 ± 0.30	0.31 ± 0.03
Hyperlipidemia					
Test I	90.8 ± 8.5	54.1 ± 2.7	148.7 ± 6.4	8.81 ± 0.68	0.29 ± 0.02
II	67.3 ± 6.3	46.6 ± 4.5	130.1 ± 10.4	10.71 ± 0.69	0.39[a] ± 0.02
Control I	84.1 ± 7.4	50.7 ± 2.1	164.0 ± 11.1	10.57 ± 1.40	0.33 ± 0.03
II	58.0 ± 7.1	45.6 ± 1.2	129.0 ± 17.0	8.83 ± 0.73	0.35 ± 0.03
Hypertension disease patients					
Test I	76.5 ± 6.4	43.1 ± 1.0	173.5 ± 2.3	5.29 ± 0.50	0.36 ± 0.03
II	63.5 ± 5.9	42.0 ± 1.2	139.3 ± 10.4	6.86 ± 0.58	0.37 ± 0.02
Control I	75.3 ± 8.1	46.3 ± 1.8	157.3 ± 10.2	5.12 ± 0.42	0.35 ± 0.03
II	63.8 ± 6.6	43.0 ± 1.6	135.0 ± 14.8	6.20 ± 0.10	0.36 ± 0.03

Note: I = before test, II = after test period.

[a] $p < 0.05$.

level of protein consumption during this period if compared with the background period. Conversely, an increase occurred in urea content that is difficult to explain, because this tendency did not correspond to the structure of differences in correlation of animal and vegetable proteins between the test and control diets. It must also be noted that a statistically significant increase in uric acid concentration occurred only in patients with HLP consuming foodstuffs with soy protein, which apparently may be of interest in deciphering the cause for this phenomenon.

TABLE 11

**Excretion of Total Nitrogen and Final Products of Nitrogen Metabolism
with Urine in Patients (Mean ± SE)**

Patients		Total nitrogen (g/d)	Urea (mmol/d)	Creatinine (mmol/d)	CGI (%)
IHD	Pretest	13.8 ± 9.4	500.4 ± 22.9	14.2 ± 1.3	104.7 ± 5.9
	Week 1	10.4 ± 0.5	395.6 ± 22.5	14.3 ± 1.2	104.9 ± 7.0
	Week 2	9.8 ± 0.3	396.0 ± 26.2	14.7 ± 0.8	108.3 ± 3.4
	Week 3	12.6 ± 0.4	480.3 ± 24.1	13.6 ± 0.7	106.6 ± 7.5
	Week 4	10.2 ± 0.3	348.4 ± 17.7	12.5 ± 1.0	96.9 ± 4.7
	Week 5	9.1 ± 0.3	363.5 ± 27.5	14.3 ± 1.0	110.7 ± 5.1
HLP	Pretest	12.7 ± 0.6	498.8 ± 57.4	13.3 ± 1.0	109.1 ± 8.9
	Week 1	10.2 ± 0.4	351.7 ± 32.8	12.1 ± 1.0	99.0 ± 6.3
	Week 2	9.3 ± 0.3	347.8 ± 40.9	13.2 ± 1.1	107.8 ± 4.4
	Week 3	14.1 ± 0.5	495.3 ± 35.6	14.4 ± 1.4	110.7 ± 11.8
	Week 4	10.2 ± 0.5	326.7 ± 32.5	13.1 ± 1.8	128.9 ± 11.3
	Week 5	9.6 ± 0.4	321.2 ± 64.8	14.5 ± 1.3	111.0 ± 9.3
HD	Pretest	13.3 ± 0.4	452.4 ± 45.1	14.8 ± 0.5	105.4 ± 2.9
	Week 1	9.9 ± 0.4	280.8 ± 22.6	15.3 ± 0.4	109.0 ± 1.8
	Week 2	9.3 ± 0.4	356.3 ± 27.2	14.5 ± 1.1	103.1 ± 3.3
	Week 3	13.1 ± 0.4	486.1 ± 49.4	14.8 ± 0.4	106.8 ± 5.1
	Week 4	10.1 ± 0.4	296.6 ± 31.8	15.5 ± 0.7	111.4 ± 3.4
	Week 5	10.4 ± 0.5	407.2 ± 40.3	15.0 ± 0.3	108.1 ± 3.9

Note: CGI, Creatinine growth index; IHD, ischemic heart disease; HLP, hyperlipidemia; HD, hypertension disease.

TABLE 12

**Nitrogen Balance Status and Protein Digestibility of the Test and Control Diets
in Ischemic Heart Disease Patients (Mean ± SE)**

Indices	Background period		Test period	
	Test group	Control group	Test group	Control group
Nitrogen consumption (g/d)	18.2	18.2	13.6	12.5
Nitrogen excretion (g/d)				
In urine	13.8 ± 0.4	12.6 ± 0.4	9.8 ± 0.3	9.1 ± 0.3
In feces	3.8 ± 0.3	4.7 ± 0.4	3.3 ± 0.2	3.6 ± 0.4
Nitrogen balance (g/d)	+0.6 ± 0.8	+1.0 ± 0.8	+0.6 ± 0.6	−3.3 ± 0.8
True digestibility[a] (%)	83.3 ± 1.6	78.4 ± 2.1	81.5 ± 1.7	77.1 ± 3.4
Net protein utilization[a] (%)	21.3 ± 4.7	23.1 ± 4.7	28.1 ± 4.4	24.1 ± 6.5

[a]   While calculating the values for true digestibility and net protein utilization, the endogenic nitrogen excretion was assumed to be equal to 2.5 g/d in the urine, and 0.75 g/d in the feces.

Table 11 shows the results of study of the excretion of total nitrogen, urea, and creatinine in urine as well as creatinine growth index calculations. It follows from the data that as it was expected, excretion of total nitrogen and urea in the urine was reduced independently on the disease type and the composition of test and control diets due to variations in the levels of consumption of protein in the background and test periods. Creatinine excretion and creatinine growth index value did not change for the observation period, which, to a certain degree, proved the adequacy of dietary protein sources to meet the body amino acid requirements of the patients. There was no significant differences in the values of nitrogen balance status, protein digestibility, and its utilization by periods and groups of patients as it is confirmed by the above data (Table 12).

Thus, the use of foodstuffs with isolated soy protein in the diet therapy of patients with obesity confirmed a previously observed effect of hypocholesterolemic action only with the presence of hypercholesterolemia. In addition, these foodstuffs provided a good feeling of repletion and significant suppression of the feeling of hunger under the conditions of reduced energy consumption that to a significant extent can guarantee patients' ability to observe a hypocaloric diet for a long period of time. The above shows perspective in the use of such types of combined products in the prevention and treatment of obesity and hypercholesterolemia.

REFERENCES

1. **Volgarev, M. N., Vysotsky, V. G., Meshcheryakova, V. A., Yatsyshina, T. A., and Steinke, F. H.,** Evaluation of isolated soy protein foods in weight reduction with obese hypercholesterolemic and normo-cholesterolemic obese individuals, *Nutr. Rep. Int.,* 39, 61, 1989.
2. **Borsoskaya, D. P. and Kavenskaya, S. D.,** Prothrombin blood activity: method of determination, *Clin. Med.,* 8, 88, 1948.
3. Fibrinolysis, in *Clinical Application of Fibrinolysis,* Andreyenko, G. V., Ed., Moscow, 1967, 214.
4. **Pokrovsky, A. A.,** *Biochemical Methods of Research in Clinics,* Moscow, 1969, 102.

Chapter 23

OVERVIEW AND SUMMARY

Fred H. Steinke, Michael N. Volgarev, Doyle H. Waggle, and Andrei N. Bogatyrev

The current diet pattern in developed countries has resulted in and contributed to the development of a number of chronic diseases which can be reduced through the implementation of healthier food products. These diseases include heart disease, whose occurrence is related to the excess consumption of high-fat, high-calorie, and high-cholesterol diets based on an increased consumption of animal foods containing these products. Numerous scientific committees have reviewed these diseases and recommended diet modification. Studies in this book have summarized the benefits of modifying foods to reduce the intake of fats and cholesterol. These objectives have been accomplished through the use of high-quality isolated soy proteins as well as other dietary modifications. Reducing the saturated fats in the diet and total fat to a more reasonable level does effectively reduce elevated cholesterol and thereby the risk of heart disease. The use of vegetable proteins in these foods will also improve the economics of food supplied to the consumer and therefore have a twofold benefit in reducing food costs and food production investment.

The papers presented in this book have reviewed both the needs and opportunities of new food protein sources to provide more economical, healthy, and nutritious foods for the population. New vegetable protein sources offer a wide variety of protein and amino acid contents which may have greater or lesser value in finished food. Certainly, vegetable proteins of high purity and functionality offer the best opportunity and the greatest flexibility in developing new food products that are adaptable to existing food systems. A number of vegetable protein sources are being tested and developed for use in the food industry. These include the soybean, pea protein, sunflower, cottonseed, and rapeseed as the primary potential new protein sources. Of these protein sources it is apparent that soybean protein is by far the most commercially advanced and has already obtained wide usage in the food industry.

Extensive nutritional work has been presented in this book summarizing the extensive studies on isolated soy protein with both animal and human clinical studies. These studies have demonstrated that isolated soy proteins can be used as the primary protein source in food products and diets. They also can be used in combination with other proteins to improve the protein value of the foods and to extend existing food sources while optimizing food quality. These studies have included infants, children, and adults which in all cases demonstrated that nutritious, good-quality food products can be provided to the consumer.

It is the hope of the editors of this book that the readers will use the information provided by the many research reports to implement development of new food products and increase the availability of them to the consumer. In particular, the development of healthy foods in many forms and for all eating patterns would be a major contribution to the health, well-being, and quality of life for the populations in both the developed and the underdeveloped world. In particular, it is hoped that there can be a shift from the heavy use of animal foods to foods which are more dependent on vegetable products having lower calories and fat content. This does not mean we recommend the elimination of animal products, but rather the rational and sensible balance of animal products and vegetable products to optimize their acceptability while providing high-quality, nutritious, healthy foods.

Continued effort should be made to evaluate the effect of diet and health through carefully conducted research projects. In conjunction with this is a need for continued development of improved food products that can be used to implement the already known health benefits that can be derived by modifying foods, menus, and eating patterns of the general populations and specific individuals at high risk of diet-influenced diseases.

Immediate and direct cooperation between science, medicine, and industry should be initiated to achieve these changes in eating patterns. This will lead to improved health and productivity of the population as well as reduced expenditure on medical care.

Programs should be initiated to include:

1. Development of a wide range of food products which positively modify dietary intake fats and calories while maintaining traditional eating quality
2. Rapid industrialization and introduction of these products to consumers
3. Development and introduction of an easily understandable public education and information system on the diet modification which clearly communicates the value and benefits to consumers
4. Follow-up studies to demonstrate to consumers the achieved results and reinforce sustained dietary change

Rapid introduction of the above must begin now in order to ensure that current and future generations adopt an improved diet and reduce the high levels of diet-related chronic diseases.

INDEX